QL681.A973 T95 1993
Tyler, Ronnie C.,
    Audubon's great
    national work

DISCARDED

COLORADO MOUNTAIN COLLEGE

Steamboat Campus
Library
Bristol Hall
1275 Crawford Avenue
Steamboat Springs
CO 80487

DEMCO

Audubon's Great National Work

Audubon's Great

National Work

The Royal Octavo Edition of
The Birds of America

BY RON TYLER

UNIVERSITY OF TEXAS PRESS · AUSTIN

Publication of the color plates has been assisted by a gift from Nelda C. Stark.

Copyright © 1993 by Ron Tyler
All rights reserved
Printed in the United States of America
First University of Texas Press edition, 1993

The paper used in this publication meets the minimum requirements of American National Standard for Information Sciences—Permanence of Paper for Printed Library Materials, ANSI Z39.48-1984.

Library of Congress Cataloging-in-Publication Data

Tyler, Ronnie C., date
    Audubon's great national work : the royal octavo edition of Birds of America / Ron Tyler. — 1st ed.
      p. cm.
    Includes bibliographical references (p.   ) and index.
    ISBN 0-292-78129-6
    1. Audubon, John James, 1785–1851. Birds of America. I. Title. II. Title: Royal octavo edition of Birds of America.
QL681.A973T95   1993
598.2973 — dc20   92-30793

Contents

List of Illustrations	ix
Acknowledgments	xiii
Methodology	xvii
1 The "Great Work"	1
2 Completion of the Double Elephant Folio	33
3 The Royal Octavo Edition	47
4 A Great National Work	73
5 Subsequent Editions of *The Birds of America*	106
6 Audubon in American Art	131
Tables	153
Appendix: Royal Octavo Editions of *The Birds of America*	163
Notes	167
Bibliography	191
Index	205

List of Illustrations

1. John James Audubon, *Self Portrait*, 1826.	frontispiece
2. John James Audubon, *Greater Prairie Chicken*, 1824.	17
3. John James Audubon, *Black-billed Cuckoo*, 1822.	17
4. John James Audubon, *Carolina Paroquet*, 1825.	18
5. Frederick Cruickshank, *John James Audubon*, 1834.	19
6. Frederick Cruickshank, *Lucy Bakewell Audubon*, 1834.	19
7. Frederick Cruickshank, *Victor Gifford Audubon*, 1836.	19
8. Frederick Cruickshank, *John Woodhouse Audubon*, 1836.	19
9. Alexander Rider after John James Audubon, *Great Crow Blackbird*, 1826.	20
10. Childs and Inman after John James Audubon, *Marsh Hens*, 1832.	20
11. Alexander Lawson after Alexander Wilson, *Mississippi Kite, Tennessee Warbler, Kentucky Warbler and Prairie Warbler*, 1811.	21
12. Thomas Bewick, *Lesser Fauvette*, 1797.	22
13. William Swainson, *Arctomys Franklinii*.	22
14. William Yarrell, *The House-Sparrow*, 1837.	23
15. J. B. Chevalier after J. Delorme, *Californian Vulture*, 1839.	23
16. The first page of the prospectus for *The Birds of America*, 1827.	24
17. Robert Havell, Jr., after John James Audubon, *Louisiana Water Thrush*, 1828.	41
18. Robert Havell, Jr., after John James Audubon, *Black-throated Guillemot*, 1838.	42

19. Robert Havell, Jr., after John James Audubon, *Black, or Surf Duck*, 1836. ... 42
20. John James Audubon, *Great White Heron*, 1832. ... 43
21. Robert Havell, Jr., after John James Audubon, *Great Blue Heron*, 1834. ... 44
22. R. Trembly after John James Audubon, *Fork-tailed Flycatcher*, 1840. ... 61
23. J. C. after John James Audubon, *American Coot*, 1842. ... 62
24. R. Trembly after John James Audubon, *Least Flycatcher*, 1844. ... 63
25. Robert Havell, Jr., after John James Audubon, *Canada Jay* (composite plate), 1838. ... 64
26. Robert Havell, Jr., after John James Audubon, *Canada Jay*, 1831. ... 64
27. Robert Havell, Jr., after John James Audubon, *Canada Jay*, 1838. ... 65
28. R. Trembly after John James Audubon, *Canada Jay*, 1841. ... 65
29. *The Camera Lucida*, undated. ... 66
30. Louis Prang, *Lithographer*, 1874. ... 67
31. Cover of Audubon's royal octavo edition of *The Birds of America*, 1839. ... 68
32. R. Trembly after John James Audubon, *Mourning Ground Warbler*, 1840. ... 85
33. W. after John James Audubon, *Black-Throated Wax Wing*, 1842. ... 86
34. John James Audubon, "Birds of America Day Book," page 120, 1841. ... 87
35. R. Trembly after John James Audubon, *Hemlock Warbler*, 1840. ... 88
36. John Woodhouse Audubon (attrib.) after John James Audubon, *Hemlock Warbler*, 1840. ... 89
37. John James Audubon, *Hermit Wood-Warbler*, 1840. ... 90
38. John James Audubon, *Willow Ptarmigan*, 1840. ... 91
39. J. C. after John James Audubon, *Willow Ptarmigan*, 1842. ... 91
40. John James Audubon, *Welcome Partridge*, 1842. ... 92
41. J. C. after John James Audubon, *Welcome Partridge*, 1842. ... 92

42. John Woodhouse Audubon after John James Audubon, *Ivory Gull*, 1843. — 93
43. John James Audubon, *Sora Rail*, 1842. — 93
44. Robert Havell, Jr., after John James Audubon, *Maria's Woodpecker. Three-toed Woodpecker. Phillips Woodpecker. Canadian Woodpecker. Harris's Woodpecker. Audubon's Woodpecker*, 1838. — 94
45. J. C. after John James Audubon, *Phillips Woodpecker*, 1842. — 95
46. Unknown artist after John James Audubon, proof of *Common Mocking Bird*, 1841. — 96
47. Unknown artist after John James Audubon, *Common Mocking Bird*, 1841. — 97
48. R. Trembly after John James Audubon, *Arkansaw Flycatcher*, 1840. — 98
49. R. Trembly after John James Audubon, *Arkansaw Flycatcher*, 1856. — 99
50. John James Audubon, *Missouri Meadow Lark*, 1844. — 100
51. Unknown photographer, *John Cassin*, c. 1860. — 117
52. R. Trembly after John James Audubon, *Swainson's Swamp Warbler*, 1856. — 118
53. Tint block for *Swainson's Swamp Warbler*, 1856. — 119
54. R. Trembly after John James Audubon, *Californian Turkey Vulture*, 1839. — 120
55. R. Trembly after John James Audubon, *Californian Turkey Vulture*, 1859. — 121
56. R. Trembly after John James Audubon, *Burrowing Day-Owl*, 1840. — 122
57. Unknown artist after John Woodhouse Audubon, *Burrowing Day-Owl*, 1859 or later. — 123
58. John Woodhouse Audubon, *14 Miles N.W. of Altar*, 1849. — 124
59. Unknown artist after John James Audubon, *Fish Hawk*, 1890. — 125
60. Robert Havell, Jr., after John James Audubon, *Great-footed Hawk*, 1827. — 141

61. Robert Havell, Jr., after John James Audubon, *Blue Jay*, 1831. 142

62. Robert Havell, Jr., after John James Audubon, *Louisiana Heron*, 1834. 143

63. George Catlin, *Buffalo Bull's Back Fat, Head Chief, Blood Tribe*, 1832. 144

64. Robert Havell, Jr., after John James Audubon, *Ruffed Grouse*, 1828. 145

65. Robert Havell, Jr., after John James Audubon, *Mourning Dove*, 1827. 146

66. John James Audubon, *Common American Wild Cat*, 1845. 147

67. Francis D'Avignon, *John James Audubon*, 1850. 148

Acknowledgments

I trace my awareness of John James Audubon and his stunning bird prints back more than twenty-five years, to my appointment as Curator of History at the Amon Carter Museum in Fort Worth, Texas. The museum possesses a unique collection of double elephant folio engravings, the first fifteen plates of *The Birds of America*, which I had the pleasure of including in several exhibitions. In researching these prints, which were among the first Audubon prints to arrive in America, I learned of Audubon's subsequent production of the royal octavo edition and finally acquired a set for the museum in 1985 so we could include them in an exhibition of lithographs relating to Texas the following year.

I noted then the lack of published information about the octavo edition but thought nothing more about it until several years later when Tom Taylor suggested that I write an essay on Audubon, perhaps about Audubon's trip to Texas, that he could publish in a fine press edition. Since I am engaged in writing a history of nineteenth-century lithographs related to Texas, this sounded like a good idea. The first thing I confirmed was the paucity of information about the octavo edition; then as I began to locate Audubon's financial records and letters, I learned that what information had been published was, as often as not, incorrect.

The deeper I got into the manuscripts, the more I realized that Audubon's publication and selling of the octavo edition between 1839 and 1844 is the story of one of the most beautiful, popular, and important natural history books published in America during the nineteenth century. Audubon's fame rests justifiably on the unparalleled double elephant folio, but the less expensive octavo edition brought his works—both visual and literary—to thousands more than would have seen the double elephant folio or read the *Ornithological Biography*, which accompanied it. I am convinced that it

Acknowledgments

earned his place as one of the most important American Romantics and have tried to understand his achievement in that larger context as well as suggest how and in what areas that influence might have spread.

No one can undertake a study relating to Audubon without quickly realizing the debt owed to one's predecessors, for the bibliography is vast and the misinformation rampant. While I am ultimately responsible for the content and conclusions reached herein, I would like to thank the many scholars and friends who have helped me in the endeavor. My first act was to contact Alice Ford, who has studied Audubon for decades, and she reminded me of the small drawings included in the H. Bradley Martin sale at Sotheby's the previous summer. They turned out to be the greatest discovery of this quest, for I concluded that they were made in the process of producing the first octavo edition and that perhaps as many as half of them are by Audubon himself, rather than by John Woodhouse Audubon, as had long been thought.

I did not really enter into octavo edition scholarship, however, until I met David M. Lank of Montreal, who has been studying birds and, more specifically, the octavo edition for years. He readily shared with me his conclusions as well as his plate-by-plate comparison of the first (1839–1844) and third (1859) states of each of the five hundred lithographs included in the octavo edition. His lecture, delivered at the twentieth anniversary celebration of the E. A. McIlhenny Natural History Collection at the Hill Memorial Library at Louisiana State University in October 1991, was most enlightening, leading me to make several revisions and additions to my text. I am indebted to him for his assistance and have tried to credit him in the notes as to specific suggestions. Still, I want to thank him here for his generous sharing of interest and information and for reading and offering suggestions and corrections to an earlier draft of the manuscript.

Special thanks go to William S. Reese, an Audubon aficionado for years, who generously provided guest quarters in his book-studded offices in New Haven while I worked in the Beinecke Rare Book and Manuscript Library at Yale. He permitted me to study his set of the *Viviparous Quadrupeds of North America*, which Audubon produced simultaneously with the octavo *Birds*, and made dozens of volumes of Auduboniana available after the Beinecke had closed its doors each day. He also read an early draft of the manuscript and made many welcome suggestions for corrections and further research. George Miles, Curator of the Western Americana Collection, and Vincent Giroud, Curator of Modern Books and Manuscripts, at the Beinecke were most helpful. Edward B. Kinney of the Audubon Gallery in Vienna, Virginia, made

Acknowledgments

suggestions based on his years of experience with Audubon prints, and he shared information from his collection, especially a set of the *Quadrupeds* text that has the six octavo plates bound in. Kenneth Newman of the Old Print Shop in New York; Judy and Jim Blakely and James Von Reuster of the Old Print Gallery in Washington, D.C.; Joel Oppenheimer of the Douglas Kenyon Gallery in Chicago; and many others have permitted me to browse through their prints or helped me sort out the various editions of Audubon's "miniature edition."

Generous friends made possible two brief respites away from the office, which enabled me to bring much of this huge pile of information into focus, one amid the hospitable confines of Sam'l and Carrie Arnold's condominium in Santa Fe, the other in William and Eleanor Crook's London flat, not too far from where Audubon spent much of his time while working on the double elephant folio.

George Ward of the Texas State Historical Association and Carol Clark of Amherst College read an early version of the manuscript and offered valuable suggestions and corrections, and I have profited from literally dozens of conversations with William H. Goetzmann of the University of Texas at Austin, who also read the manuscript. Bernard Reilly, Chief Curator of the Prints and Photographs Division at the Library of Congress, and Peter C. Marzio, Director of the Museum of Fine Arts in Houston, helped me in regard to nineteenth-century printers and lithographic techniques.

Florence M. Jumonville, Head Librarian of the Historic New Orleans Collection, assisted me with research on New Orleans subscribers to the octavo edition; Allan H. Stokes of the South Carolinian Library at the University of South Carolina provided photocopies of Bachman's reviews in the *Southern Cabinet*; and Nelda C. Stark of the Nelda C. and H. J. Lutcher Stark Foundation and Anna Jean Caffey of the Stark Museum of Art in Orange, Texas, provided access to the wonderful Audubon holdings there. At the University of Texas at Austin, Stephen C. Stappenbeck of the Barker Texas History Center assisted me with research in nineteenth-century newspapers, P. Lynn Denton of the Texas Memorial Museum provided access to the octavo editions of Audubon in their collection, and George Klos helped with research. Milan Hughston, Librarian of the Amon Carter Museum in Fort Worth, Texas, provided references and photocopies, and Mary Lampe, Coordinator of Audio Visual Materials at the Carter, made study slides of the Carter's Audubons available.

For permission to quote passages from manuscripts in their collection, I am indebted to Edward C. Carter II, Librarian of the American Philosophical Society in Philadelphia; Rodney G. Dennis, Curator of Manuscripts at the Houghton Library at Harvard

Acknowledgments

University; Jack A. Siggins, Deputy University Librarian at Yale University; and Nelda C. Stark, president of the board of the Stark Museum of Art.

A number of people have offered stylistic suggestions, including Jill Mason, who edited an early version of the manuscript; Tom Taylor, who read virtually every version of it, from its first incarnation as a twenty-five-page essay to its final form; and Paula Eyrich Tyler, my wife and my best and most insightful editor.

R. T.
Austin, Texas

Methodology

Distinguishing among the various Audubon editions and plates can be confusing because Audubon changed many names from printing to printing. Thus I have used the names that he assigned the plates rather than the current names of the birds because I am discussing Audubon's prints rather than the birds themselves. I have included plate numbers in parentheses following the title of the work and distinguish among the various publications by including Havell (for the double elephant folio), Bowen (for the octavo edition of the *Birds*, despite the fact that Endicott printed three numbers), Imperial (for the *Viviparous Quadrupeds of North America*), Octavo (for the octavo *Quadrupeds*), and Warren (to refer to B. H. Warren's *Report on the Birds of Pennsylvania*). When I refer to the original paintings, now in the collection of the New-York Historical Society, I abbreviate "O.P." and use the painting number as listed in Marshall B. Davidson, ed., *Original Water-Color Paintings by John James Audubon*. Stark (followed by a number) refers to Audubon's own set of *The Birds of America*, now in the Stark Museum of Art in Orange, Texas.

Audubon's Great National Work

CHAPTER ONE

The Great Work

More than a century and a half have passed since John James Audubon's remarkable book, the royal octavo edition of *The Birds of America*, first appeared in Philadelphia in 1839. Published serially during the next five years, this one-hundred-part set, 10½ by 7 inches (untrimmed) and usually bound in seven volumes, was based upon his more famous double elephant folio, with its life-size engravings of birds and five volumes of accompanying text, which he had published in Great Britain between 1826 and 1838. The "little work," as Audubon called it, contains five hundred hand-colored lithographic portraits, including all the birds in the larger work, plus seven species and seventeen birds that Audubon found later, and most of the text from the accompanying five-volume *Ornithological Biography*. Connoisseurs and scholars have largely shunned it, seeing it as little more than a pale imitation of the masterpiece. But the octavo edition is a significant achievement in its own right, representing the best of pre–Civil War American lithography and giving Audubon the opportunity finally to display his scholarship and genius to a large American audience for the first time.[1]

The magnificent double elephant folio of *Birds of America*, known for its brilliantly colored, life-size engravings of birds, is the largest ornithological book ever published. It contains 435 engravings, measuring 39½ by 26½ inches (untrimmed), portraying 1,065 birds (more than four hundred species), hundreds of botanical illustrations, and scores of insects, shells, and assorted detritus; the backgrounds constitute a virtual catalog of American vistas and ecological settings. Audubon painted dozens of birds unknown to science and established new standards for ornithological illustration, but his artistic accomplishment is even greater. By intuitively imbuing each image with his Romantic proclivities, he earned an as yet unrecognized place among nineteenth-century American artists. The splendid four-volume folio, which sold for approximately $1,000, immediately became an ornithological and artistic landmark, influencing forever the ways in which birds are perceived and natural history artists see the

The "Great Work"

world. "He has . . . produced . . . a work that must remain to the end of time a monument of unexampled perseverance, worthy of an ardent lover of Nature," wrote a critic for the London *Monthly Chronicle*. "It is the only work that represents birds as they are."[2]

Had the "Great Work" not been published, however, the royal octavo edition alone would surely have established Audubon's reputation in natural history art and scholarship. The unusual aspect of the octavo edition—aside from its heroic accomplishment in cataloging virtually all the birds of North America—was its financial and popular success. Appearing just as the rising tide of capitalism, literacy, and romanticism produced a large number of individuals who could and would buy such a publication, it was hailed as "a great National work." More than a thousand persons subscribed at $100 to $120, depending on the binding, one-tenth the cost of the double elephant folio but still a substantial amount, and the family reprinted it, or permitted it to be reprinted, long after Audubon's death, perhaps as many as eight times in all. Its impact upon generations of naturalists and strong appeal to a public with a long-standing love affair with the American wilderness make it arguably the single most influential work of natural history and art in the nineteenth century. Twentieth-century audiences are no different. It recently has been reproduced twice in facsimile, the illustrations themselves have been reprinted several times, and the complete edition is now available on laser-read compact disk.[3] (See Appendix.) Most people cannot tell the difference between reproductions of the double elephant folio and the octavo edition.

Despite its ubiquity, the octavo edition is probably best known today by the small, hand-colored lithographic portraits of the birds that are regularly stripped out of their bindings and sold individually in bookstores and galleries across the nation. It is no great surprise that little has been published on the octavo edition—only that Audubon scholars have not made more of it—because little has been published on the history of books in general during these years, especially illustrated books. There are some monographs on the more famous artists and several recent inquiries into the printers and printing techniques. But after the original art was created, how were the books financed and produced? How were they marketed? Who purchased such books or read them? How widespread was their influence? Because of his high public profile, voluminous correspondence, and sense of history, which dictated that he preserve all his records, Audubon himself provides some insights into these questions.

Important factors in the octavo edition's success were its beauty, comprehensiveness, and authenticity. Audubon had not conceived it or put it together in a year or even a

The "Great Work"

decade; it was the culmination of the work of a lifetime, the fulfillment of his "Great Work." Audubon's earliest memories were of birds, and his first demonstrated talent was the ability to sketch. Born on April 26, 1785, at his father's plantation on the western end of Santo Domingo (now Haiti), the illegitimate son of French navy lieutenant Jean Audubon and Jeanne Rabine, a French servant girl, he grew up in France under the doting care of his stepmother, Anne. She permitted him leisurely walks through the countryside, where he sketched birds and animals, hunted, and otherwise came to love nature, while his father, who lamented his son's lack of navigational and mathematical ability and tolerated his lack of discipline, also encouraged his artistic bent. Audubon would later claim that his father had sent him to Paris sometime during 1802–1803 to study painting with the famous Jacques-Louis David, Napoleon's court painter. This likely was but one strand of the complex web of half-truths and creative recollections that he wove around his early years throughout his life.[4]

American acquisition of Louisiana in 1803 provided Lieutenant Audubon an opportunity to send his son to the United States (with a passport declaring Louisiana as his birthplace), in part to avoid Napoleonic military conscription but also to learn English and acquire some knowledge of management and farming on his father's recently acquired Mill Grove farm, just outside Philadelphia. Audubon pursued a variety of ventures, but he never stopped studying and sketching birds. He befriended neighbor William Bakewell and courted Bakewell's daughter, Lucy, shortly after his arrival. During a return trip to France in 1805, during which the senior Audubon arranged a business partner, Ferdinand Rozier, for him, the romantic youth spent most of his time at the family villa in Couëron, near Nantes, adding to his pencil and pastel bird sketches under the tutelage of Dr. Charles D'Obigny and avoiding conscription into the army.[5]

Returning to the United States in April 1806, Audubon continued to court Lucy Bakewell, and they were married two years later. The new couple settled in Louisville, where Audubon and Rozier established a mercantile venture. There in March 1810, as Audubon later recalled, he met Alexander Wilson, the dour Scottish naturalist who was canvassing the country selling subscriptions to his handsomely illustrated and pioneering *American Ornithology*. Wilson had only two volumes of this first large-scale colorplate book printed in America to show, but the hand-colored engravings (Plate 11) fascinated Audubon to the point that he was on the verge of signing Wilson's subscription list when Rozier, knowing that he did not have the cash, asked him in French why he would make such a purchase when his own drawings were better. Audubon finally declined to purchase Wilson's book, but the ensuing conversation led him to

The "Great Work"

show the visitor his own bird paintings and drawings and spend the next few days hunting with him in the nearby woods. Following Wilson's visit, the Audubons moved on to Henderson, Kentucky, and finally to Sainte Genevieve on the Mississippi River south of Saint Louis, where the business partnership with Rozier failed. Rozier apparently wanted to live in a French-speaking settlement, while Audubon was growing accustomed to the Anglo-American frontier.[6]

Audubon was not yet ready to cast his lot with the birds, but his appreciation of Wilson's book might have inspired him to think about a giant book of his own in which the birds could be illustrated life-size. The couple returned to Henderson, and Audubon entered into a business partnership with his brother-in-law, Thomas W. Bakewell, dealing in pork, lard, and flour on commission. They also speculated in real estate for a few years, but the effects of the War of 1812, including the blockade of New Orleans, and subsequent economic depression reached the frontier just as Audubon finished construction of a sawmill that had cost more than he anticipated. The business that he expected never materialized. As the panic of 1819 forced the closing of bank after bank throughout the West, Audubon's creditors demanded payment of his note. The young couple was forced to sell everything, including their wedding silver and dinnerware, but it was not enough to pay their debts. Audubon was briefly jailed for debt in Louisville and had to file for bankruptcy. "One of the most extraordinary things among all these adverse circumstances," Audubon later recalled, "was that I never for a day gave up listening to the songs of our birds, or watching their peculiar habits, or delineating them in the best way that I could.... During my deepest troubles I frequently would wrench myself from the persons around me, and retire to some secluded part of our noble forests.... It was often necessary for me to exert my will, and compel myself to return to my fellow-beings."[7]

His continuing interest in birds and his ability to sketch made possible his next employment—as an artist and taxidermist, teaching and stuffing birds and fishes for the Western Museum in Cincinnati. It was there that Dr. Daniel Drake, founder of the museum, recognized his untutored artistic talent and gave him his first public exposure. Perhaps that was the impetus that Audubon needed to combine his two real gifts—a lifelong love of nature and a developing ability to draw—into a career. Further encouraged by a visit with Major Stephen H. Long and his party, who stopped in Cincinnati on their way back from their famous exploring expedition to the Great Plains, he was now convinced that it had fallen his lot to supersede Wilson, to document all of the birds of North America in a book that would be larger than Wilson's,

The "Great Work"

in both format and number of birds represented. "Without any Money My Talents are to be My Support and My enthusiasm my Guide in My Dificulties, the whole of which I am ready to exert to keep, and to surmount," he wrote as he embarked.[8]

Audubon's forest wanderings now became more purposeful, and the results show in his work. His early drawings, during the time he claimed to have studied with David, exhibit none of the technical knowledge or perspective that a professional artist would have imparted, but his sketches do show gradual improvement. By the time he returned to the United States in 1806, he had begun to animate his subjects and place them in realistic environments. He probably had copied the style of most contemporary natural history illustrators, hanging the dead bird by its feet or holding its head up with threads in order to draw it, but by the time he met Wilson, he had perfected his own technique, developed at Mill Grove, of using thin wires like an armature in a sculpture to animate the freshly killed birds, setting them in lifelike poses against a grid background that he had drawn on a soft board. He thus overcame the problem of how to translate onto his sketch pad the details and characteristic attitudes that he had observed in nature. An identical grid on his drawing paper helped him with the problems of perspective, foreshortening, and scale. Most of these drawings he colored with pastels, using only a few touches of watercolor.[9]

In 1820 Audubon took on the first of several assistants, a twelve-year-old boy named Joseph Mason, who had studied with him at the Cincinnati museum. They trekked down the Ohio, Mississippi, and Arkansas river valleys, rambled northeastward to Pennsylvania and upstate New York, westward to the Great Lakes, then southward to Louisiana, first in New Orleans and then at Beech Woods plantation, north of Saint Francisville, where Mrs. Audubon had, by then, taken a teaching job to provide for their young family. Audubon earned his living by doing chalk portraits, painting signs and murals for steamboats, and teaching drawing, dancing, fencing, swimming, and arithmetic wherever he could gain employment. In New Orleans, he even painted a nude portrait of a mysterious Mrs. André, for which he received an expensive "souvenir" gun worth $125. The "exuberant French coxcomb," as historian Lewis Mumford called him, had many admirers in Louisiana. Indeed, Audubon's wife, Lucy, was attracted to his natural grace and poise from first meeting him, and his handsome features and flirtations got him into trouble on more than one occasion. Dr. Louis Heermann of New Orleans dismissed him without paying for lessons that he had given Mrs. Heermann when the doctor became suspicious of the relationship. Audubon then taught dancing, drawing, and the braiding of hair jewelry at Oakley plantation, near

5

The "Great Work"

Saint Francisville, until Lucy Pirrie became distrustful of his influence over her daughter. All the while he filled his portfolio with resplendent pictures of birds and saved his money for a trip back East, where he hoped to find a publisher for his developing "Great Work."[10]

As his painterly skills developed, he began to depict more details—eyes, bills, feet—with watercolor until, by 1824, he was employing it almost exclusively, with only occasional use of the pastels that characterize his earlier work. Some of the vibrancy he depicted resulted from his working method. He sought out birds and usually observed them for days before shooting them, quickly drawing them on the spot and making extensive notes. He had long known that some of their brilliance and color faded soon after they died, and live observation ensured that his paintings were as accurate as possible. He also added lush and detailed habitats. Audubon was not the first to depict the birds in their habitat; Mark Catesby had sometimes shown them with characteristic vegetation or some plant or fruit on which they fed in his *Natural History of Carolina, Florida, and the Bahama Islands* (2 vols.; London, 1731–1748). John Bartram had included rudimentary habitats in Benjamin Smith Barton's *Elements of Botany* (Philadelphia, 1803), and Charles Willson Peale placed the specimens in habitats with painted landscape backgrounds in his Philadelphia museum. Audubon was capable of brilliant habitats and landscapes himself, as his painting of the greater prairie chicken backed with an elegant tiger lily and a western meadows scene near the Great Lakes demonstrates (Plate 2), but young Mason proved gifted at depicting flowers, branches, and habitats, which saved Audubon enormous amounts of time. After painting a bird, Audubon would select a plant or tree limb for Mason to copy, and his superb paintings helped Audubon raise the level of ornithological illustration to new heights.[11]

These years in and around Louisiana saw Audubon's greatest improvement as an artist. He seemed to draw inspiration from the woods of Bayou Sara and Saint Francisville, and his style matured. By the summer of 1822, when Audubon gave Mason a double-barreled shotgun and materials to earn his way back to Cincinnati, the teenager had added more than fifty vivid habitats to Audubon's growing collection of birds. When Audubon set out for Philadelphia in March 1824, he carried some of his and Mason's finest compositions to Philadelphia, such as the *Black-Billed Cuckoo* (Plate 3), the *Yellow-Billed Cuckoo* (O.P. 22), the *Common Grackle* (O.P. 104), and the *Great Blue Heron* (Plate 21), among the more than two hundred paintings and drawings of birds in his portfolio. He was confident that his finished works were superior to those of the "closet naturalists," who sketched their stuffed birds stiffly perched on a branch or

The "Great Work"

standing on the ground, against plain white backgrounds, from museum models that might have been improperly preserved by taxidermists who had never seen the birds alive or in their natural habitats. They were only interested in working out taxonomies for the birds, he said, in arranging "our Fauna in Squares, Circles, or Triangles."[12]

Audubon's desire for a great book indicated his ornithological as well as artistic aspirations, for illustrated books had been the favored method of transmitting scientific and technical information throughout Europe, at least since the publication of Conrad Gesner's *Historia Animalium* (4 vols.; Zurich, 1551–1558), one 800-page volume of which was devoted to descriptions of birds and illustrated with more than two hundred woodcuts. It appeared simultaneously with Pierre Belon's *Histoire de la Nature des Oyseaux* (Paris, 1555), containing approximately two hundred hand-colored woodcuts of birds, but both were eclipsed by Ulisse Aldrovandi's encyclopedic 2,600-page *Ornithologia* (3 vols.; Bologna, 1599–1603). Literally dozens of books appeared during the ensuing decades, with early explorers of the New World collecting information on birds and French artist Jaques le Moyne producing the first illustration of a North American bird in 1564 (wild turkeys, published in Theodore De Bry's *Brevis narratorio eorum quae in Florida Americae Provicia Gallis acciderunt* . . . [Frankfurt, 1591]). The first illustrated book on American birds to reach the European market was Catesby's *Natural History of Carolina, Florida, and the Bahama Islands*.[13]

Audubon did not fully understand the economic dimensions involved in producing an illustrated bird book, particularly one in which the birds would be reproduced life-size, requiring a larger book than had ever been published. Nor did he realize the social and professional jealousies that he would arouse by setting his untutored hand to such a task. Ornithology had remained popular throughout the seventeenth and eighteenth centuries, particularly among the wealthy and well-educated classes of Great Britain and Europe, and they, along with the natural history artists such as Thomas Bewick (Plate 12), John Gould, Prideaux John Selby, and others, sustained a modest industry devoted to the production of luxurious illustrated books. Barons and dukes crisscrossed the continents of the New World, seeking new and fascinating species, then shared their findings through illustrated books, which were the crowning glory of several noble expeditions into the heart of the American continent, from Baron Alexander von Humboldt's *Essai Politique sur le Royaume de la Nouvelle-Espagne* (4 vols.; Paris, 1811) and *Researches, Concerning the Institutions and Monuments of the Ancient Inhabitants of America* . . . (London, 1814) to Prince Maximilian of Wied-Neu Wied's breathtaking *Reise in das Innere Nord-Amerika in den Jahren 1832 bis 1834* (Coblenz, 1839–

1840), which was illustrated by the young Karl Bodmer's unparalleled depictions of Plains Indians. Those who could not lead the expeditions financed them and later made the trips vicariously through the grand books that resulted. A good collection of illustrated natural history books became the mark of a gentleman's library, and the authors who produced such works were lionized in the Old World and the New.[14]

Philadelphia was the cultural and publishing hub of the United States; and here Audubon expected to find the best artists, naturalists, and printers in the young country. As the American printing industry developed, William Bartram had published his *Travels Through North and South Carolina, Georgia, East and West Florida* there in 1791. William R. Birch and his son, Thomas, had published *The City of Philadelphia in the State of Pennsylvania North America* (Springland Cottage, near Bristol, Pa., 1800) and *The County Seats of the United States* (Springland, near Bristol, Pa., 1808). Scotsman Alexander Lawson, the best-known engraver in the city, had produced the plates for Wilson's nine-volume *American Ornithology; or, The Natural History of the Birds of the United States* (Philadelphia, 1808–1814), of which Audubon had seen a portion in Louisville in 1810. The greatest intellects of the young country had patronized the American Philosophical Society, the Library Company, the Academy of Natural Sciences, and Charles Willson Peale's museum. They had instructed Meriwether Lewis in preparation for his 1804–1806 expedition to the Pacific Ocean with William Clark, and Dr. Edwin James had just published Major Long's account of his expedition to the Rocky Mountains, modestly illustrated with engravings made from sketches by Samuel Seymour and Titian Ramsay Peale. Dozens more of their pictures, along with specimens of birds and animals that few but the mountain men and the Indians had seen, remained unpublicized in their collections.[15]

By the time Audubon arrived, however, Philadelphia no longer boasted geniuses of the stature of those who had established its reputation. Bartram had died the previous year. Charles Willson Peale was in his eighties and well beyond his prime; his son Rembrandt recognized the undisciplined potential in Audubon's bird portraits, but another son, Titian, who, because of his own natural history collections and drawings and hard-earned experience from the Long expedition, could have been of great assistance to Audubon, refused to help.

Dr. James Mease, whom Audubon knew from his Mill Grove days, cordially welcomed the thirty-nine-year-old artist and introduced him to the portrait painter Thomas Sully and to twenty-one-year-old Charles-Lucien-Jules-Laurent Bonaparte, the prince of Canino and Musignano (as a result of his ambitious uncle Napoleon's conquests).

Audubon had changed his buckskin hunting clothes for a new black suit, in imitation of his hero, Benjamin Franklin, but had refused to cut his shoulder-length hair, which, in his vanity, he considered to be characteristic of a frontiersman. Sully appreciated Audubon's self-tutored ability and admired his courage in seeking scientific support even though Audubon was completely unschooled in the subject. He even asked the frontiersman to instruct his daughter in the use of pastels and watercolors. Bonaparte probably inquired politely as to Audubon's French origin but was likely so confused by the reply, which included a reference to Audubon's father, the admiral, as well as the claim of studying under David, that he changed the subject to natural history. He was then at work on a supplement to Wilson and was so intrigued by Audubon's paintings and firsthand experience that he took him to a meeting of the Academy of Natural Sciences, where he showed his work to the likes of George Ord, the friend and biographer of Wilson. Ord was an influential figure who served as both secretary of the Philosophical Society and vice president of the academy. He had finished Wilson's *Ornithology* after the naturalist's death and was then revising it for reprinting in a smaller format.[16]

Audubon had not reckoned on the jealousy with which these savants guarded the academic portals—and Wilson's reputation. Ord did not take kindly to such an obvious backwoodsman claiming to have eclipsed Wilson in number of species as well as style and accuracy of drawing. Titian Peale, whom Audubon might have met as a member of the Long expedition, refused to cooperate because he was illustrating Bonaparte's four-volume *American Ornithology; or, The Natural History of Birds Inhabiting the United States, Not Given by Wilson* (Philadelphia, 1825–1833) and feared that Audubon might become a competitor for the job. The *coup de grâce* came from printmaker Alexander Lawson, the Scotsman who had engraved and owned a percentage of Wilson's book and was, at the time, working on Bonaparte's book as well as Ord's new edition of Wilson. Lawson later recalled that the prince had brought Audubon by the printing shop to show his work. Bonaparte clearly liked the paintings, and Lawson admitted that they were "extraordinary for one who is self taught," but he pointed out that they were "too soft, too much like oil painting" to be engraved. When Bonaparte put one of Audubon's paintings beside one of Wilson's, thinking that the direct and, no doubt, dramatic comparison would sway Lawson, the printmaker's comments grew sharper: Audubon's painting was "ill drawn, not true to nature, and anatomically incorrect."[17]

Undaunted, Bonaparte commissioned Audubon to paint *A Pair of Boat-Tailed Grackles* to Lawson's specifications and excitedly presented the finished painting to the printer with the request that it be engraved for his book. Lawson refused, saying, "Ornithology requires truth in forms, and correctness in the lines. Here are neither." The embarrassed Audubon attempted to defend his work—and his ego—with the false boast that he had "been instructed seven years by the greatest masters in France," but Lawson had the last word: "Then you made dam[n] bad use of your time." Lawson finally engraved the grackles, but only after Alexander Rider redrew and reduced the picture, for, as Bonaparte later explained to Audubon, the figures, although reduced from his usual life-size, were still too large to be engraved for his book. The grackles were published as the *Great Crow Blackbird*, plate 4 in Bonaparte's *American Ornithology* (Plate 9).[18]

Audubon biographers generally discount Lawson's criticisms and agree with Charles Winterfield's more positive assessment that "such delineations as Audubon's . . . are an immortal satire upon this whole school of stuffed-specimen pretenders . . . who have afflicted Science with the nightmare of their wretched figures and indigestible crudities." They also conclude that more than technique was involved in Lawson's rejection of Audubon and that the Philadelphians did not intend to give the upstart, self-taught naturalist the satisfaction of having his work appear in print. Ord went so far as to contact Audubon's former assistant, Joseph Mason, then an artist for the academy and disappointed that his former mentor had not given him credit for his part in the pictures, to obtain damaging information about Audubon's professional ethics.[19]

In all fairness, though, Lawson had the standard practices of the engraving trade on his side when he rejected Audubon's work. At the time, engravers asked artists to produce complete compositions that could be copied precisely. Because they depended upon the strong, black lines of the engraving to create the image, they did not want to be placed in the position of having to approximate the subtle tones of watercolor or oil paintings with crosshatching and multiple lines. They demanded that the strokes in the painting be made exactly as the engraver would make them on the copper, whether the process was engraving, etching, or aquatint. Texture and tone were to be indicated; shadows and perspective were to be shown. Artists felt that such requirements inhibited their artistic freedom and constantly complained, but they made the engraver's job easier and a more accurate reproduction possible.[20] Because Audubon's

The "Great Work"

work was mostly watercolor, it lacked the strong, clear lines that the engraver needed to copy. Lawson would not only have had to reduce Audubon's painting, no mean feat in itself, but decide how to re-create in line the subtle tones, shades, and forms of his vibrant watercolors. Audubon's paintings were beautiful, but to Lawson they were not adequate for engraving.

Audubon's trip was not a complete loss. In addition to Sully and Bonaparte, he met artist-naturalist Charles-Alexandre Lesueur and engraver Gideon Fairman, who had just returned from England and become a partner with Philadelphia printer Cephas G. Childs. Both suggested that Audubon go abroad to look for an engraver. A work of the magnitude that he envisioned would demand the best, Fairman said, and the supply of craftsmen and copper was more abundant in London. During a trip to New York in August, where he presented a paper to the Lyceum of Natural History and was elected to membership despite having to defend himself against charges that Ord and Lawson had passed along to their Manhattan brethren, Audubon found publishers there as unenthusiastic about his "Great Work" as the Philadelphians had been. That fall he returned to Louisiana by way of Albany, Niagara Falls, Pittsburgh, Cincinnati, and Louisville and began to prepare for a trip to Great Britain.[21]

Because he was self-taught, Audubon probably did not realize that the pattern of winning approval in England and Europe in order to be accepted at home had been *de rigueur* for American artists who had been crossing the Atlantic for two generations. Benjamin West and John Singleton Copley felt it necessary to live in London to pursue their careers, and they attracted others, such as John Trumbull, Gilbert Stuart, and Ralph Earl. For the later generation, including Thomas Sully, John Vanderlyn, and Washington Allston, the grand tour of the Continent was routine. There was good reason for this, of course. They received training from acknowledged masters in England and Europe that they could not get in the United States. Alexander Wilson did not have such certification, nor was he accepted as an artist outside of natural history circles, and it had taken him years to earn the respect of the Philadelphia mavins for his pioneering efforts.[22]

Meanwhile, Audubon continued to make huge improvements in his work, producing some of his finest compositions in 1825. The Louisiana swamps and forests yielded the *Carolina Paroquet* (Plate 4), the *Wild Turkey* cock (O.P. 1) and hen (O.P. 5), the *Ivory-Billed Woodpecker* (O.P. 181), and the *Mocking Bird* (O.P. 44, Plates 46, 47). Perhaps the assistance he needed did lie in Great Britain, where a larger community

The "Great Work"

of ornithologists and collectors would help him accomplish his goal, but the American back-country provided the raw material for his genius. The British would later see him at his Romantic and artistic best.[23]

Armed with a sheaf of introduction letters, Audubon sailed for England in May 1826. Landing in Liverpool almost two months later, the black-clad frontiersman, his "flowing curling locks" hanging down to his shoulders, went first to the office of his brother-in-law Alexander Gordon, husband of Lucy's younger sister Ann. Gordon received him "coldly" and did not even offer his home address until the artist had paid the customs duties on his drawings himself. The following day, Audubon began delivering his letters of introduction. "Mr. Audubon who carries this has at length decided upon seeking for that patronage in Europe which he cannot find here," one such note read. "I wish him success but am afraid he will not meet with that success which his drawings deserve." Others were more enthusiastic. A letter from exporter Vincent Nolte of New Orleans directed him "through the sinuous streets of Liverpool" to his first important contact, the Rathbone family, the first importers of American cotton in England, and to the praise that he had vainly sought in Philadelphia. "He carries with him a collection of upwards of 400 Drawings, which far surpass anything of the kind I have yet seen . . . and convey a far better idea of American Birds than all the stuffed birds of all the museums put together," Nolte wrote. There, in front of the Rathbone family and friends, Audubon opened his portfolio of birds. "I knew by all around me, that all was full of best taste and strong judgment, but I did not know if I would at all please," he confided to his journal, which he wrote as if it were a letter to Lucy. He need not have worried: "These friends praised my Birds, and I felt the praise, yes breathed as if some celestial being succored me in Elysium," he wrote. The Rathbones helped him arrange the first exhibition of his paintings in Liverpool.[24]

The Rathbones' support even brought Audubon's family around. On August 5, Audubon's sister-in-law Ann Gordon wrote to Euphemia Gifford, a wealthy relative in Derby, that Audubon had "brought a great many letters to persons of the first respectability in this place, and he has already received a great deal of attention on account of the beauty of his collection of birds. I wish you could see them." She added, in a postscript: "Since writing the above I have seen Mr. Audubon, who says he wishes to visit you, and I do say he will be with you soon after the receipt of this. You will find him pleasant in his manners and I hope much improved in character."[25]

Although he possessed a natural grace and style, Audubon had been in America for decades, and his plain, black suits and long hair marked him as a true frontiersman in

The "Great Work"

socially intimidating situations. He has "no dash, no glimmer or shine about him," Sir Walter Scott would later remark of their first meeting, "but great simplicity of manners and behaviour...." Audubon's healthy ego remained intact, however, and he wrote Lucy that his unshorn tresses attracted the attention of "every *lady* that I met," who "looked at them and then at me until—she could see no more." The self-portrait done at this time as a favor for Mrs. Rathbone confirms his standing as a romantic figure (Plate 1, *frontispiece*).[26]

Audubon and his birds evoked a similar response in others. Historian William Roscoe was full of "surprise and admiration" for the startling creations of one so sure of himself in the woods but apparently so ill at ease in society. The coeditor of the *Encyclopedia Britannica* and influential antiquarian book dealer Henry G. Bohn of London initially told Audubon that a smaller format book would be more suitable for the English market, perhaps starting him thinking in terms of an octavo publication. "Remember ... that productions of taste are purchased with delight by persons who receive company," he said, "and that to have your book be laid on the table as a pastime piece of entertainment is the principal use [to be] made of it." But Audubon was so determined to fulfill his dream of life-size prints that the idea of a smaller book lay fallow until he became more familiar with the market for such a work; and even Bohn, upon viewing the stunning original paintings, agreed that "they must be published the full size of life." There was a further reason: Wilson's *American Ornithology* had been reprinted in New York and Philadelphia, and some copies were being distributed in England, so Audubon needed a competitive edge against a book already on the market. A trip to Edinburgh, Scotland, and a meeting with one of Britain's foremost engravers, William H. Lizars, were even more rewarding: "My God," Audubon quoted the printmaker as saying, "I never saw anything like this before!" Audubon considered his prompt election to membership in the Wernerian Society of Natural History at the University of Edinburgh just recompense for his insulting rejection in Philadelphia. "I am in miniature in Liverpool what Lafayette was with us," he wrote his son Victor soon after his arrival. "My exhibition attracts the *beau monde*."[27] Benjamin Franklin charming the literati of Paris might have been a more apt comparison.

It was during a subsequent exhibition in Manchester that the American consul Joshua S. Brooks, a native of Boston, suggested that Audubon sell his proposed "The Birds of America" by subscription and issue it serially, the method Wilson had used for his *Ornithology* and one common in England and on the Continent. Nineteenth-century, large-format, illustrated books were most often issued over a number of years in fascicules, or parts, each containing five or more illustrations and usually some descriptive letterpress text. A paper cover bearing the author's name, the part number, the date of publication, and sometimes the prospectus of the work along with a list of subscribers completed the package. Needless to say, such lengthy projects were confusing as some subscribers dropped out and others were added, leaving some original subscribers with only a portion of the set and often requiring the author to reprint earlier numbers to supply his new customers with the entire edition. The authors marketed their books by publishing prospectuses and pursuing as many potential purchasers as possible through personal contacts, natural history societies, and advertising (through the press and free of charge, if possible). The king or queen, in Britain or on the Continent, was the first target and, if royalty subscribed, the book would then be offered to others with their "particular patronage, approbation and protection." The author and his cohorts would plan to sell a certain number in Britain and on the Continent and in Audubon's case, in America; usually about two hundred subscriptions were required to ensure the economic success of such an endeavor. Such books were so expensive that there was virtually no other method available for financing them, so Audubon adopted the plan. One of his first letters was to Charles-Lucien Bonaparte. Others soon followed to De Witt Clinton, Henry Clay, Andrew Jackson, and William Clark, all of whom had supplied him with letters of introduction, but none of whom subscribed.[28]

In Edinburgh Audubon had first called on the eminent Scottish naturalist, Robert Jameson, who was engaged with Prideaux John Selby in the production of *Illustrations of British Ornithology* (2 vols.; Edinburgh, 1833). But it was Patrick Neill, a printer and horticulturist, who introduced him to Lizars, the foremost printmaker in Edinburgh and the engraver of Selby's work as well as the British edition of Wilson's *American Ornithology*. The originality and quality of Audubon's paintings immediately overcame any misgivings that Lizars might have had, and he agreed to engrave Audubon's birds. Audubon then prepared a prospectus (Plate 16) announcing a publication of eighty numbers, each to contain five prints of double elephant folio size (39½ by 26½ inches), so the birds could be represented in life-size, and priced at two guineas (£2.2, or approximately $10 at the 1828–1838 rate of exchange).[29]

Audubon contracted with Lizars for fifty impressions of each of the five engravings in the first number. He would order more copies as he sold subscriptions. Lizars presented him the proof of plate 1, the *Wild Turkey*, on November 26, 1826, and the other prints for the first number followed rapidly. "Mr. Audubon," Lizars reportedly said,

The "Great Work"

"the people here don't know who you are at all, but depend upon it they shall know." By May 1827, Audubon had ordered fifty more copies of the first number and had instructed Lizars to print one hundred copies of number two. He estimated that he would only cover expenses until he reached two hundred subscriptions, at which time the profit on each number would be £174.18.8 sterling (approximately $775). If Lizars could get out five numbers per year, Audubon estimated, the project would take sixteen years to complete and his profit would be £879.7.4 sterling, or $3,902 per year—"enough to maintain us even in this country in a style of Elegance and Comfort," he wrote to Lucy.[30]

Audubon, meanwhile, was lionized in Scotland. He met with naturalists and professors, dined at society's table, was elected to the various natural history societies, and delivered formal papers on the habits of alligators and the rattlesnake (which he confused with a blacksnake in his painting of the mockingbird). He also demonstrated his technique of wiring up birds in front of a grid to the members of the Wernerian Society. He was the first bird painter to attempt anything more than zoological draftsmanship. His method permitted him to depict birds in dramatic poses, with wings spread and tails fanned. His insistence upon including natural habitats, while not unique, was novel, and his later expansion into full-fledged landscapes made his pictures much more interesting to the landed gentry in both Britain and Europe—but especially in the United States, where curiosity about the continental landscape was growing and the first American school of landscape painting—the Hudson River School—was developing. The most unusual aspects of his paintings were their aesthetic quality and vibrancy, which created a sensation among naturalists and artists alike. Upon viewing Audubon's watercolors on exhibition at the Royal Society, a well-known French critic wrote of their "magic power" by which they "transported us into the forests which for so many years this man of genius has trod." Audubon finally received a coveted invitation to call on Sir Walter Scott, perhaps the most famous man in Britain, who recalled him as "plainly dressed; wears long hair which time has not yet tinged; his countenance acute, handsome, and interesting."[31]

With production of his book underway, Audubon now turned to other tasks. First was the sale of enough oil copies of his bird and animal paintings to provide for his immediate survival. While watercolors were adequate for the engraver, collectors treasured oil paintings, perhaps for their permanence as opposed to the fragility of watercolors, but Audubon never mastered the medium. The chance acquaintance of a nineteen-year-old artist, Joseph Bartholomew Kidd, whom Lizars had employed to fill in the sky behind one of Audubon's bird paintings, seemed to offer a partial solution: Kidd agreed to copy Audubon's birds in oil, both for sale and for a proposed traveling exhibition by which Audubon hoped to earn enough money to support production of the *Birds*.[32]

With Lizars's engravers at work on the second number of the "Great Work," Audubon prepared to go to London, then the center of the natural history world, by sacrificing his cherished curls in the hope of making himself more presentable to the cognoscenti. Most of the subscription sales would have to be accomplished there. He set out in April 1827, with copies of his prospectus and the first five engravings. He paused at Newcastle upon Tyne to visit with engraver Thomas Bewick and sold a subscription to the Philosophical Society of York. In Leeds he sold five, in Manchester, eighteen more. He renewed his acquaintances in Liverpool, especially with the Rathbones, then finally arrived in London, where John George Children, the retiring secretary of the British Museum's Department of Antiquities and former editor of two scientific journals, cordially received him. Children, who would prove to be an important friend, arranged for an exhibition of the birds at the Linnaean and Royal societies. Another important contact was the painter Sir Thomas Lawrence, a colleague of Sully's, who offered to help Audubon find buyers for the oil copies of his bird paintings as well as to teach him to paint and glaze oils. King George IV's subscription crowned Audubon's efforts in September 1827.[33]

Meanwhile, the whole project was threatened in June when Lizars informed Audubon that his colorists had gone on strike and that work had stopped. Lizars had attempted to find others, even taking on some students, but without success. Only the first two numbers had been printed, and not all of them colored. He suggested that Audubon might want to take the prints on hand and find colorers in London.[34]

Audubon fell into depression and concluded that he had to find another printer immediately or the project would be doomed. He might have gone to Paris, as several had recommended, because it was widely assumed that the best engravers and colorists in Europe could be found there. The young German-Russian artist Ludovic Choris had gone there in 1818 with his watercolors from Otto von Kotzebue's around-the-world voyage, as would Karl Bodmer when he returned from his North American adventure in 1834. Instead, Audubon approached London engraver Robert Havell, whose shop at 79 Newman Street he regularly passed on the way to and from his rooms on Great Russell Street. Havell had attracted Audubon's attention because he also operated a natural history gallery there, dealing in such items as bird skins and

2. John James Audubon, *Greater Prairie Chicken*, 1824. Watercolor, 25 by 36 in. O.P. 55. Courtesy The New-York Historical Society. Audubon, the consummate natural history artist, painted the elegant tiger lily and the meadow in the background, as well as the birds.

3. John James Audubon, *Black-Billed Cuckoo*, 1822. Pencil, watercolor, pastel, and lacquer, 19 3/8 by 24 1/8 in. O.P. 32. Courtesy The New-York Historical Society. Audubon created some of his best paintings with the assistance of the teenage Joseph Mason of Cincinnati. Audubon would paint the birds, then select a shrub, flower, or limb for Mason to copy. The *Black-Billed Cuckoo* is one of their best collaborations.

4. John James Audubon, *Carolina Paroquet*, 1825. Pencil, watercolor, and lacquer, 29 3/4 by 21 1/4 in. O.P. 223. Courtesy The New-York Historical Society. The Philadelphians who rejected Audubon's work did not see some of his best pictures. He continued to improve after his return to Louisiana and painted these Carolina parakeets near Bayou Sara, Louisiana. The birds were declining in Audubon's day and are now extinct.

5. Frederick Cruickshank, *John James Audubon*, 1834. Miniature on ivory. Private collection.

6. Frederick Cruickshank, *Lucy Bakewell Audubon*, 1834. Miniature on ivory, original no longer exists. Courtesy The Charleston Museum, Charleston, South Carolina.

7. Frederick Cruickshank, *Victor Gifford Audubon*, 1836. Miniature on ivory. Private collection.

8. Frederick Cruickshank, *John Woodhouse Audubon*, 1836. Miniature on ivory. Private collection.

"Cruikshank [sic] has Improved my Miniature very considerably—he has worked over the Hair &c—This picture goes to [C.] *Turner* to be engraved in Mezzotints," Audubon wrote Bachman on August 25, 1834. Havell printed the engraving early in 1835. Cruickshank also painted portraits of the other family members, but Lucy's was destroyed in a fire.

19

9. Alexander Rider after John James Audubon, *Great Crow Blackbird*, 1826. Hand-colored engraving by William Lizars, 9 3/8 by 12 1/2 in. (sheet). From Bonaparte, *American Ornithology*, plate 4. Courtesy Stark Museum of Art, Orange, Texas. Audubon's first published picture appeared in Bonaparte's *American Ornithology*, but not until Alexander Rider redrew it according to Lawson's specifications. Audubon was furious when he saw this published version.

10. Childs and Inman after John James Audubon, *Marsh Hens*, 1832. Lithograph, 19 3/4 by 26 1/4 in. (sheet). Courtesy Stark Museum of Art, Orange, Texas. While Audubon was in Philadelphia in 1832, he had this lithograph made, perhaps in an effort to decide if he might be able to do the "small work" by lithography.

11. Alexander Lawson after Alexander Wilson, *Mississippi Kite, Tennessee Warbler, Kentucky Warbler and Prairie Warbler*, 1811. Hand-colored engraving, 13 1/8 by 10 in. (sheet). From Wilson, *American Ornithology*, Vol. 3, plate 25. Courtesy Stark Museum of Art, Orange, Texas. Wilson put several birds on a page, not to illustrate both sexes of a species or to show relationships, but to show as many birds as possible in a limited amount of space. When Audubon first saw Wilson's book in Henderson, Kentucky, he probably began thinking about doing a bird book of his own. He concluded, however, that he would show only one species per plate.

12. Thomas Bewick, *Lesser Fauvette*. Woodcut, 3 1/8 by 2 1/4 in. (image). From Bewick, *Birds of Great Britain* (1797, 1804). Private collection. Seeing the beauty and success of Bewick's books on the birds of Great Britain—the nearest thing to a popular guide to birds at that time—probably inspired Audubon to pursue a "miniature edition" of his own.

13. William Swainson, *Arctomys (Spermophilus) Franklinii*. Etching, 8 1/4 by 11 in. From John Richardson, *Fauna Boreali-Americana*...(1829-37). Courtesy Department of Special Collections, University of Chicago Library. Audubon admired Swainson's style and carefully studied his work before undertaking the octavo edition.

14. William Yarrell, *The House-Sparrow*, 1837. Engraving, 2 3/4 by 2 3/16 in. (image). From Yarrell, *A History of British Birds* (1837-43), Vol. 1, 470. Courtesy Jeff Weber Rare Books. Audubon feared that the works of Yarrell, John Gould, and others would satiate the market for small bird books.

15. J. B. Chevalier after J. Delorme, *Californian Vulture*, 1839. Hand-colored lithograph, 10 1/4 by 6 5/8 in. From John Kirk Townsend, *Ornithology of the United States of America...* (Philadelphia, 1939). Courtesy Rare Books Division, Special Collections, University of Virginia Library. Another threat might have come from Dr. John Kirk Townsend, who invited Audubon to coauthor a book on American ornithology with him. Audubon declined, and Townsend's book failed to find a market. It is rare today, possibly because only a few books were published as a trial to see if it would sell.

BIRDS OF AMERICA,

FROM

DRAWINGS

MADE

DURING A RESIDENCE OF UPWARDS OF TWENTY-FIVE YEARS

IN

THE UNITED STATES AND ITS TERRITORIES,

BY

JOHN JAMES AUDUBON,

CITIZEN OF THE UNITED STATES;

MEMBER OF THE LYCEUM OF NEW YORK; FELLOW OF THE ROYAL SOCIETY OF EDINBURGH; MEMBER OF THE WERNERIAN NATURAL HISTORY SOCIETY; FELLOW OF THE SOCIETY OF SCOTTISH ANTIQUARIES AND NEW CASTLE TYNE; MEMBER OF THE SOCIETY FOR PROMOTING THE USEFUL ARTS IN SCOTLAND; OF THE LITERARY AND PHILOSOPHICAL SOCIETY OF LIVERPOOL, &c. &c.

Prospectus.

To those who have not seen any portion of the Author's splendid collection of original Drawings, it may be proper to explain, that their superiority consists in every specimen being of the full size of life, pourtrayed with a degree of accuracy as to proportion and outline, the result of peculiar means discovered and employed by the Author, and lately exhibited to a meeting of the Wernerian Society. Besides, in every instance where a difference of plumage exists between the sexes, both the Male and the Female Birds have been represented. The Author has not contented himself with single profile views of the originals, but in very many instances he has grouped them, as it were, at their natural avocations, in all sorts of attitudes, either on branches of trees, or amidst plants or flowers: some are seen pursuing with avidity their prey through the air, or searching diligently their food amongst the fragrant foliage; whilst others of an aquatic nature swim, wade, or glide over their allotted element. The Insects, Reptiles, or Fishes, that form the food of the birds, have been introduced into the drawings; and the Nests of the Birds have been frequently represented. The Plants are all copied from nature, and the Botanist, it is hoped, will look upon them with delight. The Eggs of most of the species will appear in the course of publication.

16. The first page of the prospectus for *The Birds of America*, probably sent out in May 1827. Courtesy American Philosophical Society, Philadelphia, Pennsylvania.

The "Great Work"

shells, and he was well known for his delightful portfolios of colored aquatint engravings, *A Series of Picturesque Views of the River Thames* (London, 1812) and *A Series of Picturesque Views of Noblemen's and Gentlemen's Seats, with Historical and Descriptive Accounts of Each Subject* (London, 1823). The fifty-eight-year-old engraver immediately recognized the boldness of Audubon's project but declined the job because of his age; he reasoned one as old as he should not undertake a task that seemed certain to require more than a decade and a half to complete.[35]

Then fate seemed to intervene on Audubon's behalf. Soon after his visit, a friend, Paul Colnaghi of Colnaghi and Company, called on Havell to show him the work of one of his young engravers. Havell liked the work, and with Audubon in mind, he asked to meet the promising craftsman. Colnaghi is said to have replied, "Then, send for your own son." Havell and his son, Robert, Jr., had quarreled over whether the younger Havell would be permitted to be an artist and engraver rather than pursue one of the professions, as his father wanted. Thirty-three-year-old Robert, Jr., had taken a sketching holiday in Monmouthshire and was earning his living by obtaining commissions from various publishers, Colnaghi among them. Havell and his son reconciled, formed the partnership of Robert Havell and Son, and agreed to take on the project, with the younger Havell handling the engraving and aquatint and his father supervising the coloring staff, which at the height of the project would number fifty men and women.

Although Lizars had lost some of his enthusiasm for Audubon's project as a result of the difficulties, he offered to share it with the Havells so as to retain a portion of the business. Audubon, on the other hand, despite Lizars's advice and personal kindness, was relieved to turn both the printing and coloring over to the Havells, who had bid twenty-five percent less than Lizars had charged (£115, or approximately $510, for one hundred copies of each number of five prints each). The senior Havell printed and colored many of the plates himself before his death in 1832, while his son produced the immense double elephant–sized copperplates, some measuring more than six square feet, by etching, aquatint, and engraving.[36] It is possible, of course, that Audubon might have changed engravers anyway, for Havell was a better printer than Lizars. Still, he could have done so in a more orderly manner if given the choice.

The transition from Lizars to the Havells went smoothly enough, but a group of prints, among the first to reach American shores (and now in the collection of the Amon Carter Museum in Fort Worth, Texas) illustrates some of the minor difficulties Audubon had to overcome. Audubon sent instructions to Lizars by Kidd, a chance

25

visitor in London, to ship all copies of the first two numbers—colored and uncolored—to London. When Audubon saw them, he realized that Lizars had misnamed the *Yellow-Billed Cuckoo* (Havell 2), calling it the *Black-Billed Cuckoo* and giving the Latin name for that species, along with several other small errors. A few incorrectly labeled prints of the *Cuckoo* found their way into circulation, but Audubon had Havell re-engrave the copperplate with the correct title on subsequent printings. On at least one copy, now in the Carter collection, Havell simply erased the incorrect title and penned in the correct one in a fine Spencerian hand. There are similar variations in others of the Lizars prints that Havell later printed.[37]

Audubon was, in fact, in the hands of an as yet unrecognized master. As his birds began to pour from the press, Robert Havell, Jr., rapidly became known as the best engraver in Britain. Engraving, despite the fact that there is a precise meaning for the term—a metal plate incised by a graver—was often used, even in the nineteenth century, to refer to a number of intaglio processes. Historically, it has included engraving (in its true sense), etching, aquatint, mezzotint, and drypoint, among others. Audubon's prints are almost always referred to as "engravings" because both Lizars and Havell credited themselves with engraving—"Engraved Printed & Coloured by R. Havell"—on the prints, but they may be more accurately identified as etchings and aquatints, with the precise technique of engraving sometimes used only if it suited Havell's purpose. Etching and aquatint, combined with Havell's superior coloring, produced what are most likely the most beautiful bird prints ever made, a considerable improvement even over Lizars's impressive work.[38]

Audubon could now turn his attention to other matters because production of the prints was in good hands. His appreciation of Bewick's books, plus Bohn's recommendation of a smaller format, led him back to the idea of an octavo edition of the *Birds* late in 1827. Because of the trouble with Lizars immediately after he arrived in London, it was some time before Audubon was able to express himself on the subject. Then he couched it, not in the context of the further flowering of an idea, but in terms of the economic soundness of the path that he had chosen. In December 1827, in the course of assuring Lucy that he would succeed and that she would be able to join him the following year, he wrote, "As to publishing Birds, I am now well aware that I can undertake different *Small* works with immense advantage as I know much of the tricks of the trade."[39]

Audubon laid any additional projects aside in September 1828, however, as he prepared for his delayed trip to Paris in quest of subscriptions. There the Baron Georges

The "Great Work"

Léopold Chrétien Frédéric Dagobert Cuvier, preeminent naturalist and zoologist of the duke of Orléans's Museum of Natural History, was so taken with his prints that he made an immediate appointment to show them to the duke, who promptly signed his name to Audubon's list of subscribers. (In addition to wanting famous autographs, Audubon apparently thought that the signatures on his subscription list constituted a legally binding agreement for the purchase of the entire set.) The great flower painter Pierre Joseph Redouté appreciated Audubon's work sufficiently to exchange prints from his *Belles Fleurs* for Audubon's *Birds*. And at a meeting of the French Royal Academy of Sciences, Cuvier pronounced Audubon's work "the greatest monument yet erected by Art to Nature."[40]

The success of the *Birds* was, nevertheless, still in doubt when Audubon packed for America on April 1, 1829. He was homesick and yearned to see his family, but he also had business matters on his mind. Havell had notified him that without an additional supply of bird paintings, soon "I shall be standing still for work." Audubon also knew that he had to sell more subscriptions, or he would not be able to pay the mounting printing bills. "I wish to employ and devote every moment of my sojourn in America at Drawing such Birds and Plants as I think necessary to enable me to give my publication *throughout* the degree of perfection that *I am told* exists in that portion already published and now before the Public," he wrote his wife. Leaving his paintings in care of his new friend Children, and with the first fifty of the Havell prints in a splendid new binding to show his countrymen, he booked passage to New York. He had thought of traveling incognito as "Mr. John James" so that his English subscribers would not know he had left the country and conclude that his publication would not be completed. Children, however, assured him that absence for such valid reasons would have no impact upon them.[41]

New York harbor was a welcome sight early in May, but Audubon received anything but an enthusiastic greeting. Dr. Thomas P. Jones had published Audubon's article on rattlesnakes in the *Journal of the Franklin Institute* in July 1828, which brought immediate protests, one calling it "a tissue of the grossest falsehoods." Jones had to admit in the next issue (August, 1828) that the domestic chaos of his move from Philadelphia to Washington to assume editorship of the journal had led to "hastily selected" articles and that he had read Audubon's article neither in manuscript nor in proof. In New York, meanwhile, Audubon's "immense book" had "been laid on the table of the Lyceum and look*d* at [and] praised," but the reception was little more than polite. After receiving only one enthusiastic review and selling no subscriptions, he gathered his prints and crossed the Hudson to Camden, New Jersey, to lick his wounds and go birding again.[42]

The next several days he spent watching a family of vireos build their nests and hatch their eggs at Great Egg Harbor, and his bitterness faded. Nature had restored his perspective again. Following a month on the Atlantic Coast and two weeks in Philadelphia, where he employed landscape painter George Lehman, a former partner in the lithographic firm of Childs and Lehman, to provide the backgrounds for several of his birds, he spent August in the Great Pine Forest near Mauch Chunk, Pennsylvania. He returned to Camden in the fall at the finish of a remarkably productive six months; he had painted ninety-five superb watercolors of birds and sixty different eggs. He had intended to include the eggs in his book but ultimately was unable to do so.[43]

Audubon wanted his family to return to England with him and had instructed Lucy to join him in Philadelphia so he would not have to leave the East Coast and be further out of touch with England and news of the *Birds*. But she refused. His three-year absence in England had strained their relationship even more than had his wanderings of the previous decade. Twenty-year-old Victor Gifford and John Woodhouse, now seventeen, did not want to go to England, nor was Mrs. Audubon altogether sure of her wishes. She had wanted to join Audubon soon after his arrival in Edinburgh, but he had refused, fearing that the project might fail. Then, after the *Birds* seemed off to a good start, he wrote for the family to join him, but Lucy answered with a caution that he interpreted as indifference. Now she was stably situated at Saint Francisville and did not know what fate awaited her in England. Recognizing his wife's reluctance, Audubon altered his plans and went west to gather his family. He rejoined his sons in Kentucky and asked their help, then continued to Louisiana to convince Lucy to accompany him back to England. After their reunion in Louisiana, she agreed to return with him, and he spent the next few weeks happily roaming the familiar woods and swamps in search of new birds. He added two to his total and made several more paintings before they departed in early January 1830.[44]

Audubon stopped in Washington en route to call on President Andrew Jackson and sold a subscription to the Library of Congress. In Baltimore he signed three collectors to subscriptions, including Robert Gilmor, a merchant and significant patron of American artists. Audubon had learned, meanwhile, that things were not going well in London. Children was too ill to offer Havell any real help, and there were letters of complaint and cancellation, some of his agents having roiled Havell and subscribers alike. "Keep up a good heart," Audubon advised Havell. "We will be in London as soon as possible. I will carry with me some Drawings that I know will make the *graver* and the *Acid* Grin again."[45]

Upon his return, Audubon quickly concluded that his absence was not the only problem. After exhibiting some of Havell's new prints in Manchester in June, he realized that the quality of the coloring varied from print to print. "Pay the strickest [sic] attention to the Colourers," he admonished Havell. "I could have had many new names [subscribers] at Manchester, had not the people there seen different setts in different Houses *almost of different colours* for the same plate.—I myself saw 2 setts that I scarcely could believe had been sent from your House." Hearing more complaints in Birmingham, Audubon concluded that he bore part of the responsibility for the problem—"I have examined you may depend at great leisure all the Plates as they hang on the walls around me, and I am surprised myself to see how carelessly *I have past [sic] over faults*"—but demanded that Havell produce more consistent work. After seeing proofs of plate 19, he was "much pleased with them all" and concluded that "you have either improved very much or have been more careful than ever before as I think these the best of your productions." "I am delighted with the plate of the 3 owls [*Little Screech Owl*, Havell 97]," he wrote on November 30. "It is truly beautiful." As he grew more accustomed to working with Havell, he began to employ some shortcuts, asking Havell to combine existing paintings in a couple of instances (*Barred Owl*, Havell 46, for example) to produce the finished engraving.[46] Havell's accomplishments under these circumstances, bearing in mind the difficulties Lawson had outlined, are truly remarkable.

Audubon had not originally planned a text to accompany the engravings, for his unlettered comments were the result of long and intense observations rather than a scientific understanding, as his papers before the Wernernian Society and in the *Journal of the Franklin Institute* and the subsequent furor showed. The days of the gentleman naturalist were numbered; he was being replaced by trained specialists on the Continent as well as in Great Britain and America. Still, Audubon became convinced of the necessity of a text—all such books contained one—and of the value of his notebooks, primary sources that were full of observations from which he drew his insights. "I know that I am not a scholar," he began, but "with the assistance of my old journals and memorandum-books, which were written on the spot, I can at least put down plain truths, which may be useful, and perhaps interesting." To protect himself against further blunders, he enlisted professional assistance in the person of William Mac-

The "Great Work"

Gillivray of Edinburgh, a talented naturalist and an able writer and teacher. Thus began a collaboration that lasted until the final volume of Audubon's *Ornithological Biography* was printed in 1839. The volumes were published separately from the plates and in an octavo format. Two years were required to produce the first one; five ultimately appeared, providing an articulate and readable commentary on *Birds of America* in the order of their publication.[47]

Audubon published the text separately from the prints because, if the folios had contained text, he would have been obligated under the British Copyright Act of 1709 to deposit a copy in each of nine libraries throughout the United Kingdom—an expensive prospect. He feared that even the addition of title and contents pages would make him liable under the law, so ultimately, he distributed only a title page to his subscribers. (Readers have benefited from his decision, for double elephant folio pages full of type would have been difficult to read.) Audubon ordered 750 copies of volume one to be printed by Patrick Neill of Edinburgh. He initially planned to charge subscribers one guinea ($5) per volume of text, twenty-five shillings (almost $6) for nonsubscribers, but after delivering the first volume to a subscriber, he decided against charging them. At the same time, he ordered 500 copies of volume one printed in Philadelphia in order to secure an American copyright on the text. Audubon printed considerably more copies of the *Ornithological Biography* than of *Birds of America*, the best estimate of the latter being about 176 complete copies.[48]

As Havell worked on, meanwhile, the number of Audubon's subscribers, who came mostly from the English aristocracy, continued to decline from 180 to 130. King George IV died in 1830; King Charles X of France, also a subscriber, was driven from the throne in the same year and fled to England; others simply dropped their subscriptions. Audubon again turned, for birds and subscribers, to America. "It is my intention," he wrote to Cambridge professor John Stevens Henslow upon departing again for the United States in 1831, "to give a last ransacking of the woods with a hope to find something new."[49]

Audubon's second return was a triumph. News of his project had spread. *Blackwood's* and the *Literary Gazette* welcomed him with favorable reviews. Enough of his work had been seen to attract followers, and he was finally acknowledged as a celebrity in his adopted country. Congressman Edward Everett, whom he had met in Boston in 1830, arranged exhibitions at the Library of Congress and the Boston Athenaeum; he also nominated Audubon for membership in the American Academy of Arts and Sciences. He even distributed prospectuses to potential subscribers. Finally,

The "Great Work"

the American Philosophical Society, the most prominent naturalist group in the country, subscribed and admitted him to membership in July.[50]

For the next three years, Audubon traveled widely and collected dozens of new specimens. Henry Ward, his taxidermist, and artist George Lehman accompanied him to the South, first to Charleston, South Carolina, then to Florida, where they spent the fall of 1831. Back in Charleston, a hotbed of naturalists ever since English naturalist Mark Catesby had been there, Audubon stayed with the Reverend John Bachman, a Lutheran minister and amateur naturalist who took an interest in his work and insisted that he use the Bachman home as headquarters while in the area. Maria Martin, Bachman's sister-in-law and a talented painter of brightly colored flowers and insects who would later become the Reverend's second wife, began to assist Audubon by painting backgrounds for his birds. Other naturalist friends of Bachman supported Audubon and appreciated his accomplishments. One individual, in addition to the Charleston Library, the Natural History Society, and the Citizens' Library, subscribed to the double elephant folio. Audubon paid grateful tribute to the city by including Lehman's pristine view of its distinctive skyline, with the well-known Castle Pinckney, a Revolutionary War harbor fort, in the left background of his harmonious picture of the *Long-Billed Curlew* (O.P. 182), painted that fall.[51] All in all, it is one of his most successful and popular compositions.

In April 1832, Audubon again visited Florida, this time its east coast, in search of sea and shore birds. He paused in Charleston in June as he and son John followed the migratory birds to their summer homes. In Philadelphia he published a print of *Marsh Hens* (Plate 10) with Childs and Inman, an experiment in lithography possibly made with an eye toward a future small edition of the *Birds*. He moved on to Boston, where his enthusiasm for painting a prize specimen golden eagle that he obtained from the New England Museum almost cost him his life. Hoping to preserve its brilliant plumage, he tried to suffocate the bird with charcoal gas, but it stood unfazed and refused to succumb while he and John repeatedly left the room for fresh air. Audubon finally killed the bird with a knife and spent the next sixty hours painting its portrait before collapsing with a "spasmodic affection" that required the care of three physicians before he recovered.[52]

He traveled north to Labrador in June 1833 to study the summer plumage and breeding habits of the water birds. He also wanted to see the West and secured a promise from Secretary of War Lewis Cass that if the government organized an expedition to the Rocky Mountains, he would be appointed the naturalist. However, he was not ready to make the trip until he had finished the *Birds*, for he turned down a later opportunity to go with his friend Edward Harris, a gentleman farmer and naturalist from Moorestown, New Jersey, saying that he would need to do so before he undertook the "miniature edition," but not until then.[53]

As Audubon prepared to return to England, he had no sense of his place among the first generation of American artists who were helping to define a national consciousness based on the country's long and influential relationship with nature. Naturalists and authors like Bartram, Wilson, and Michel Guillaume Jean de Crèvecoeur had prepared the way for William Cullen Bryant and James Fenimore Cooper. Artist Thomas Cole personified wild nature as America's greatest natural and moral resource and in so doing, raised landscape painting to a par with history painting. He is generally credited with fathering the country's first homegrown painting movement, the Hudson River School. The Peales and genre artists John Quidor and William Sidney Mount added their own unique visions. American art was coming of age, and William Dunlap, a portraitist turned historian and self-appointed chronicler of American art, set out to describe its energy, quality, and achievements in his two-volume critique, *History of the Rise and Progress of the Arts of Design in the United States* (New York, 1834). Even though Audubon's bird portraits seemed to place him outside the artistic mainstream, he could hardly be avoided in such a survey. "I saw the plates...admired them generally, some of them much," Dunlap wrote, "and I admired the energy he had shown, so far in accomplishing his purpose." But then he recorded Alexander Lawson's version of his encounter with Audubon in Philadelphia, leaving the impression that, while there might be some contribution to knowledge in the new species that Audubon had documented, there was little in the way of art or design that should be considered.[54] Perhaps Dunlap himself did not realize the contribution that Audubon was making because it was outside the academic tradition that he so valued.

The popular response to Audubon would have been a better gauge. "*The Natives* are quite astonished at my Production and collections &c.," Audubon had written from Philadelphia in the summer of 1832. And Philip Hone, the mayor of New York and an artist in his own right, concluded that Audubon's *Birds of America* "is probably the most splendid book ever published." The whole of Audubon's work, in fact, placed him squarely in the universe of American Romantics—authors as well as artists—but it would be some time before the critics realized this.[55]

CHAPTER TWO

Completion of the Double Elephant Folio

Even as Audubon rushed to complete the double elephant folio, the planned small edition was never far from his mind. Sometime after publication of the *Marsh Hens* in Philadelphia, he wrote his son Victor, who had remained in London to oversee production of the double elephant folio, that one possibility for the "reduced size Edition" was simply to cut down the margins of the small plates (approximately 19 by 12 inches) that Havell was producing for *The Birds of America* and use them. They could be "retouched and better finished" and would be of such quality that they might not have to be colored at all. Only the "middle and large plates" would have to be redone completely. Not only would this result in quite a savings, but the small birds would still be "given the size of nature." In addition, the edition could be produced soon after the double elephant folio was finished. The text could be included with the plates, as Bewick, Wilson, Bonaparte, and others had done. Another possibility Audubon mentioned was a smaller size, such as Bewick used for his wood engravings. All the prints would have to be redone in this instance, but the smaller size would again result in considerable savings in production costs and paper.

There would be a further advantage. The double elephant folio was so expensive that few copies would be sold to anyone "besides Public Institutions and Men of great Wealth.... A reduced Edition would be within almost every person's compass, and I believe would meet with a great Sale both in Europe and in our Country," Audubon wrote to Victor. "I still look upon such an Edition," he continued, "as on the greatest profits we are to derive from our Publications."[1]

Audubon always feared that someone else would publish a small edition before he could get to it. One piracy had already occurred: In 1831 Captain Thomas Brown began publication of what one author called an "Audubonized edition" of Wilson and Bonaparte; that is, he changed the positions of the birds and provided new back-

The Double Elephant Folio

grounds of characteristic shrubs, trees, and flowers. The work was completed in 1835, but the few copies that exist today suggest that few were printed, and the venture probably was a commercial failure. Kidd, who had copied a number of Audubon's paintings might also have been a threat—he is credited with four of the plates in Brown's publication—so Audubon cautioned Victor that they must keep their discussions secret and act quickly. He wanted neither copyists nor purchasers, or possible purchasers, of the "Great Work" to know that a less expensive edition might be forthcoming.[2]

"*If we can spare the funds* for such an undertaking my opinion is, that the money . . . could not be better employed," he wrote. He admonished Victor to look deeply into the matter and have estimates made. Should the plates be engraved on wood, steel, or copper? Should they be the size of Havell's small plates, "Bewick's Birds," or "*The Zoological Gardens*"? How many copies should be printed, and where could the work be best produced? France, Germany, or England? "This I would leave to your taste, prudence or wishes," he told Victor—but only to a point: "Perhaps indeed Bewick's Size would prove the best if the Wood Cuts were equally good as his, cheaper than any other and consequently more profitable to us all." Once all these questions had been resolved, Audubon advised, "Seek out a trusty Engraver who would undertake the whole of the Engraving, and who would go on with it at once, one possessed of such talents as would answer to our wishes in the style of the workmanship—yourself taking out of his hands into your possession each plate as soon as approved by yourself and kept in good order and Insured." Apparently Audubon planned to have the plates ready but would not print the edition until the double elephant folio was completed. "In a word," he concluded, "act on this with the same prudence, care and taste which you have so well shown, since you have been at the wheel in England."[3] At this point, Audubon seems not to have considered lithography at all. Perhaps he was not satisfied with the print of the *Marsh Hens*, which was produced by lithography in Philadelphia.

Upon reflection, Audubon realized that a publication relying upon the actual Havell plates for the small birds would still be larger than Wilson's book and several times the size of Bewick's. And, it would remain expensive. Still not sure of the best format but apparently favoring the smaller size, he continued to rehearse the options for Victor:

Perhaps *excellent Wood cuts* on the plan of Bewicks birds would prove most profitable to us—The Drawings would have to [be] very correctly diminished and exactly copied in their details, Specific characters well kept &c &c—to spend

The Double Elephant Folio

whatever money can be spared to have the Plates, either Steel, Copper or Wood prepared and finished as opportunity and *money* will admit of—perfectly secretly—taking the plates away from the Engraver into your possession when *you* are satisfied with them, and have the whole so managed that this little Edition might come before the Public when the great Work is completed, at a time most agreable to ourselves all.—the letter press attached to the plates as in Bewick *&c.*—

"This Edition" he concluded, "would contain many valuable additions in the Letter press [illegible] take well with the trade and I think pay us a very handsome amount[?]." Havell, who was by now producing prints that even Audubon considered superior, was the printer of choice: "I cannot believe that *he* will give up such [a] publication," Audubon reasoned, "for he must know that without it, he must suddenly fall into the background as an Engraver of such works."[4] Little did Audubon realize that engraving itself was falling into the background.

Audubon left Victor to follow up on his suggestions, but continued to turn ideas for a small edition over in his mind. "I have been *examining closely* the Etchings of quadrupeds by *Thomas Landseer* in Richardson & Swainson work [Plate 13], and thinking the style beautiful," he wrote Victor in November. "I think that it would be well to speak *to him* respecting the Engraving *or Etchings* of the Plates for the petite Edition of which I have spoken to you—do this with caution however." The next month he inquired if Victor had any news of the "Little Edition."[5]

By May 1834, Audubon was back in England and at work on volumes three and four of the double elephant folio. The enterprise had become a genuine family effort. John and Victor had taken painting lessons so they could be of more help, with Victor working on the landscapes and backgrounds and supervising Havell, and John, the better artist, painting birds and plants. "My work is now progressing fast," the senior Audubon advised a patron, "having made every arrangement within my power to have it finished in the course of three years, having added Engravers and Colourers to our stock of workmen for the purpose." The end of his "Great Work" was now in sight. He announced that the last number of the *Birds* and the final volume of text would appear in 1838.[6]

As he waited for volume three of the *Ornithological Biography* to be printed in December 1835, almost ten years after publication of the first number of the *Birds*, Audubon returned to the idea of a small edition. Perhaps engravings would be too expensive, but he did not really want to consider MacGillivray's suggestion of an even less expensive edition, illustrated with uncolored woodcuts. He wrote Richard Rathbone, soliciting the names of good lithographers who might undertake the small edition.[7]

When Audubon returned to the United States with John in August 1836, it was to finish his announced task of depicting all the birds of America. His greatest lacuna was the western birds. He had not traveled beyond Louisiana and had illustrated only those western birds whose skins he had been able to obtain from explorers and correspondents. Thus he was excited to learn of naturalists Thomas Nuttall's and Dr. John Kirk Townsend's expedition to the Pacific Northwest with fur trader Nathaniel Wyeth in 1834 and called on the Philadelphia Academy of Science, which held the collections, hoping to be permitted to draw the new specimens. Told that he would have to obtain Nuttall's permission, Audubon made a trip to Boston to visit the industrious Harvard professor, who had already explored the Arkansas territory and the Ozarks as well as the Northwest and had published volumes on land and water birds as well as important botanical texts. The two found that they had much in common, and Audubon returned to New York triumphant: he had purchased ninety-three duplicate skins from Nuttall. He was only slightly disappointed when several of them proved to be of birds he had already drawn.[8]

Temporarily satisfied in his quest for western birds but unable to visit Florida again because the Seminole war was still in progress, Audubon accepted the Reverend Bachman's advice to look again to Texas, which had recently won its independence from Mexico with a surprising victory at San Jacinto, near present-day Houston. It was an area that he knew to be important in the bird life of the continent, and he had tried unsuccessfully to gain employment as naturalist and draftsman on the team surveying the Louisiana-Texas boundary in March 1821. Although he was aware that the members of the Long expedition, whom he had met in Cincinnati in 1820, had noted the presence of wild turkeys and other birds along the Canadian River, in what is today the Texas Panhandle, he did not know of Swiss botanist Jean Luis Berlandier's extensive collections in South Texas as a member of the Mexican boundary survey in 1828–1829 under General Manuel de Mier y Terán, or of the Scot Thomas Drummond's collections, gathered during a twenty-one months' sojourn in south central and southeast Texas in 1833–1834. Furthermore, Audubon could have saved himself the trip if he had seen Mary Austin Holley's 1836 observations. Author of the first book on Texas in English, she accurately noted, at least for the settled portions of the state that she had seen, that, "the principal specimens of the feathered tribe in Texas . . . are, mostly, such as are well known in the United States, and therefore need no description."[9]

The Double Elephant Folio

Some stability had returned to Texas in the little more than a year that had passed since Sam Houston and his ragged Texas army defeated Santa Anna's Mexican troops at San Jacinto. Settlers who had fled the Mexican army's advance were still returning home, and would-be immigrants who might have paused to be sure that the region was safely in Anglo-American hands had resumed their migration. Through the efforts of his friend, James Grimshaw, Audubon obtained the service of a revenue cutter, as he informed another friend, Edward Harris, so that "we ... may proceed towards the Sabine River, all intermediate places, and ... even to Galveston Bay! This Bay abounds with Birds of rare plumage." The American navy had swept the Mexican blockaders from the Texas coast only a few days before the Audubon party, composed of the naturalist, his son John, and Edward Harris, arrived—simultaneously with the British vice-consul from Tampico, whom Audubon supposed to be on a "secret mission."[10]

By this time, Audubon was known even in Texas, and his arrival on April 24, 1837, was duly noted in the first issue of the *Telegraph and Texas Register* that was published in Houston: "Audubon is one of the very few Americans whose fame has extended throughout the civilized world and whose services have commanded a national tribute of respect from the United States," the editor wrote. On April 29, Audubon's son John made what might have been the first sketch of the village of Galveston with his camera lucida, an optical device that seems to project an external view or object onto the artist's paper so that it can be traced. More important, however, father and son saw dozens of birds on the island.[11]

But the highlight of the trip was meeting a bird of a different kind—"the Raven"— President Sam Houston of Texas. After meeting Audubon in Galveston, Secretary of the Navy S. Rhoads Fisher arranged an appointment for the naturalist in the new village of Houston. Audubon first glimpsed the president as he emerged from one of the city's saloons in his effort to prevent the sale of liquor to Indians. He "wore a large gray coarse hat, and ... is upwards of six feet high, and strong in proportion. But I observed a scowl in the expression of his eyes that was forbidding and disagreeable." Later, in the president's cabin, Fisher ushered Audubon into "what in other countries would be called the antechamber.... The ground-floor ... was muddy and filthy," Audubon recalled. "A large fire was burning, a small table covered with paper and writing materials, was in the centre, camp-beds, trunks, and different materials were strewed around the room." First, he met the other members of Houston's cabinet, only "some of whom" seemed intellectually up to their task.[12] By the time Audubon was summoned into his presence, Houston had changed for entertaining and "was dressed in a fancy velvet coat, and trowsers trimmed with broad gold lace; around his neck was tied a cravat somewhat in the style of seventy-six." President Houston was hospitable and offered drinks to his guests. He introduced the members of the cabinet and staff and invited Audubon to make his needs known. "Our talk was short," Audubon remembered, "but the impression which was made on my mind at the time by himself, his officers, and his place of abode, can never be forgotten."[13]

Audubon recognized Texas as one of the best ornithological observation points on the continent. "More than two-thirds of our species occur there," he estimated, but ultimately his findings only echoed those of Holley. "We found not one new species," he told his friend Thomas M. Brewer of Boston, editor of the *Boston Daily Atlas*, "but the mass of observations that we have gathered connected with the ornithology of our country has, I think, never been surpassed. I feel myself now tolerably competent to give an essay on the geographical distribution of the feathered tribes of our dear country."[14]

A number of authors have suggested that Audubon painted several bird portraits while in Texas. The image most commonly attributed to the state is the *Spotted Sandpiper*, which has a delightful scene of a gently flowing creek and pasture encircled by a Virginia worm, or zig-zag, fence made of split rails, once common throughout Texas, in the background. One author even went so far as to identify the creek as being Buffalo Bayou, most of which is today the Houston Ship Channel. Portraits of the whooping crane and other well-known Texas species are also commonly attributed to the trip. Audubon saw these birds in Texas, but his well-known painting of the spotted sandpiper was made years before, in August 1821, near Bayou Sara, Louisiana. He painted the whooping crane three months later near New Orleans.[15] None of the species in the double elephant folio *Birds of America* is, in fact, directly attributable to a Texas specimen, or to a skin obtained in Texas. Audubon had time to include portraits of Texas birds, but because he found no new species, there was no need to interrupt his already crowded work schedule to paint a Texas specimen. He did, however, make extensive use of his Texas observations in the *Ornithological Biography*.[16]

Audubon had intended to return overland, but the hardships of overland travel probably changed his mind, and he departed Galveston by sea on May 18. He paused in Charleston for son John to marry Maria, the daughter of the Reverend Bachman; then he moved up the East Coast trying, during the depression year of 1837, to collect monies due him for subscriptions.[17]

In August Audubon returned to England and to a complicated task. He found production on the *Birds of America* proceeding smoothly, with numbers 72 through 76

The Double Elephant Folio

(Havell plates 356 through 380) issued that month. Because he could not get the 489 species that he had discovered into the eighty numbers that he had promised his patrons, he extended the series by seven numbers and began crowding several birds and species into a single engraving in order to represent them all in the allotted 435 plates. Havell completed numbers 77 through 87 (Havell 381 through 435) after Audubon's return and again came to his rescue by working from a number of unfinished originals. Audubon's painting of *Townsend's Warbler* (O.P. 186), for example, contains four different kinds of warblers and two different kinds of bluebirds, eleven birds in all. Havell distributed them over two plates, 393 and 394, one with five birds, the other with six. Twelve other paintings were so complicated that Havell simplified the image or, as in the case of the *Black-Throated Green Warbler* (O.P. 327), again made two plates, numbers 399 and 414. The 435 plates contain 1,065 individual birds. One authority estimated that Audubon had missed only about 16 birds in the eastern United States that a "capable birdwatcher" would have seen.[18]

The largest birds were a little cramped even on the spacious double elephant folio sheet, but Audubon's self-imposed requirement for life-size illustration required some truly graceful and perhaps Orientally inspired solutions, as in the *Wood Ibiss* (Havell 216), the *Louisiana Heron* (Plate 62), the *American Flamingo* (Havell 431), and the *Great Blue Heron* (Plate 21), for example. Audubon's accomplishment is, perhaps, best seen by comparing his great blue heron with Catesby's. Catesby's heron does not relate to the space he occupies; he seems confined and belittled by it. Audubon's, on the other hand, is caught in that moment just after he has landed. He folds his wings as he steps forward, eyes downward, to begin the quest for "finny prey." He exudes enormous energy as he fills the entire picture—indeed, his beak ventures beyond the confines of the composition's edge, and he seems about to step out of the image itself.

The size was ideal for the medium-sized birds, such as the ducks, permitting both sexes to be shown. Here, again, Audubon's sense of drama and composition raised his portraits above the usual. An otherwise routine portrait of the *Black, or Surf Duck* (Plate 19) is given an extraordinary energy by Audubon's placing them on and next to a rock, permitting the negative space between them to restate, in a startling and forceful manner, the female's profile. The same is true of the *Tropic Bird* (Havell 262), the *Bonapartian Gull* (Havell 324), and the *Arctic Yager* (Havell 267). Audubon pictured them against the sky, a negative and sometimes dramatic space, and permitted it to echo and react to their forms, shapes, and movement in ways almost as interesting as the birds themselves.

This left Audubon with the problem of how to handle the huge leftover space in the plates of the smaller birds. He solved it by providing elaborate habitats, grouping several members of the same species or several species in some of the plates, and mixing his elegant, tiny birds with luxurious leaves and flowers. He was aware of his achievement, writing Lucy that "the little drawings in the center of those beautiful large sheets have a fine effect and an air of richness and wealth that cannot [but] help insure success." The first number that Lizars printed showed this versatility, and Audubon followed the practice of illustrating a large bird in the first plate of each number throughout the project.[19]

Audubon's was a collaborative achievement, organized in a manner similar to the old master studios. Otherwise, he could never have finished. Between 1831 and 1838 he and his assistants—his sons along with Maria Martin and George Lehman—produced half of the watercolor illustrations in *Birds of America*. At the same time, Joseph Kidd produced perhaps one hundred oil copies of birds and animals. This increased output required Havell to supply more complex parts of the pictures because Audubon did not have time to complete all the compositions. The 1832 print of the *Northern Goshawk* (Havell 141) is, perhaps, the worst example of their collaboration in a situation where Audubon requested much of him. To save time, Audubon had cut two of the birds, the adult goshawk (left) and Cooper's hawk (right), out of his earlier pastel drawings and had pasted them on the same page with an immature goshawk that he had painted in 1830. He left it to Havell to provide the background landscape. Audubon did not want his subscribers to know that he worked in this manner, of course, and he advised Victor that, "as many birds have been *Pasted*, take great care of those Drawings and shew them to *a very few* of your friends." In this instance, however, he paid a severe price, for Havell confused the perspective and the resulting image must be the worst composition in the entire book.[20]

Fortunately, there are better examples of Havell's contributions. In the plates of the *Turkey Buzzard* (Havell 151) and the *Yellow-Crown Warbler* (Havell 153), both finished in 1832, Audubon asked only that he finish out the painting of a branch in the first instance and two small limbs in the second. But in the plate of the *Barn Owl* (Havell 171), done in 1833, Havell added a peaceful, evening landscape. By 1835 the printer was routinely adding backgrounds to Audubon's birds (see Havell 265 of the *Buff Breasted Sandpiper* and 277 of the *Hutchins's Barnacle Goose*), sometimes inappropriately, as in the case of the *Noddy Tern* (Havell 275), in which the background does not match the one that Audubon later described in his comments on the bird and its habitat. One of his most fantastic is the ice landscape background for the *Black-throated Guillemot*

17. Robert Havell, Jr., after John James Audubon, *Louisiana Water Thrush*, 1828. Hand-colored aquatint and engraving, 19 3/4 by 12 1/2 in. (plate). From *The Birds of America*, plate 19. Courtesy Stark Museum of Art, Orange, Texas. Audubon painted the Louisiana water thrush near Bayou Sara in September 1821. Mason contributed the handsome Jack-in-the-pulpit with fruit.

18. Robert Havell, Jr., after John James Audubon, *Black-throated Guillemot*, 1838. Hand-colored aquatint and engraving, 18 5/8 by 28 in. (plate). From *The Birds of America*, plate 402. Courtesy Stark Museum of Art, Orange, Texas. Behind the crested auklet, ancient murrelet with young, least auklet, and rhinoceros auklet, Havell placed a fantastic landscape that makes a more pleasing picture than the one he did for the *Goshawk*, but it is no more realistic. Havell had never seen the Labrador coast and did not know what it looked like.

19. Robert Havell, Jr., after John James Audubon, *Black, or Surf Duck*, 1836. Hand-colored aquatint and engraving, 21 1/4 by 30 1/4 in. (plate). From *The Birds of America*, plate 317. Courtesy Stark Museum of Art, Orange, Texas. Audubon strengthened this rather humble portrait of the surf scoter by echoing the profile of the female (right) in the outline of the rock and the male perched on it.

20. John James Audubon, *Great White Heron*, 1832. Watercolor, 25 5/8 by 38 3/4 in. O.P. 219. Courtesy The New-York Historical Society. Audubon provided Havell instructions as to how to finish this painting, and Lehman made the landscape painting of Key West, Florida.

21. Robert Havell, Jr., after John James Audubon, *Great Blue Heron*, 1834. Hand-colored aquatint and engraving, 38 1/8 by 25 1/2 in. (plate). From *The Birds of America*, plate 211. Courtesy Nelda C. Stark, Orange, Texas. Audubon crowded the life-size portraits of the largest birds onto the double elephant folio, ingeniously contorting them in awkward but apparently natural poses to make them fit.

The Double Elephant Folio

(Plate 18). Audubon often gave Havell careful instructions, as in the case of the *Great White Heron* (Plate 20), painted at Key West in 1832 and printed in 1835 (Havell 281): "Keep closely to the Sky in depth & colouring!" he wrote in the foreground of the painting. "Have the water a *Pea-green* tint. Keep the division of the scales on the leg & feet white in your engraving—the colouring over these will subdue them enough!" Lower in the foreground, he penned: "Finish the houses better from the original which *you* have," referring to Lehman's landscape that had been sent separately. Then, in the lower right-hand corner, he added, "Have the upper back portion only mellowing in the outline."[21] The final print shows that Havell paid close attention to his client's instructions.

Indeed, Havell's feat in engraving, printing, and coloring the *Birds* is a major achievement in itself. When Audubon began his project, engraving was the preferred method of illustrating bird books, and he worked with the two best in Britain, Lizars and Havell. But by the time he finished, most engravers and aquatinters had gone out of business, and even craftsmen of the stature of Lizars and Havell had little work, because the faster and less expensive process of lithography had come to dominate the business of reproducing pictures. In fact, Audubon's was the last job that Havell did in England. In September 1839, he sold his business and with his wife and daughter, moved to New York, where he earned his living primarily by painting, with some engraving on the side. When Lizars died in 1859, the craft of using metal engravings and aquatints to reproduce bird illustrations died with him. Audubon's *Birds of America* was the last and greatest illustrated bird book to be produced by this technique in Great Britain.[22]

As soon as Havell had pulled the last proof on June 20, 1838, Audubon began work on the last two volumes of the *Ornithological Biography*, making information from his extensive Texas observations available to the naturalist world for the first time. Then he finished *Synopsis of the Birds of North America*, an index to the *Birds* and the *Ornithological Biography*, which came out a year later.[23]

Although Audubon received some criticism, particularly from his Philadelphia colleagues, there were contemporary critics who recognized his accomplishment. "In my estimation, not more than three painters ever lived who could draw a bird," William Swainson had written in Loudon's *Magazine of Natural History* as the project began. "Of these the lamented Barrabaud [Barraband], of whom France may be justly proud, was the chief. He has long passed away; but his mantle has at length been recovered in the forests of America." An anonymous London critic praised the completion of Audubon's "magnificent undertaking" in 1839, while on the other side of the

45

The Double Elephant Folio

Atlantic, a *New York Mirror* writer welcomed him home: "Audubon is a man of genius, and he has done more for the science of ornithology than any man since the days of Wilson. He deserves well at the hands of this country." The *North American Review* called *Birds of America* his "imperishable monument." Audubon had an immediate influence on several British natural history artists, including John Gould, Sir William Jardine, Edward Lear, the German-born Josef Wolf, and Prideaux John Selby, who was at work on his *Illustrations of British Ornithology* when Audubon arrived in Edinburgh and later took lessons from the American. These artists, following his example, painted birds life-size and greatly enhanced their compositions by depicting them in flight, in dramatic poses, and in natural habitats. "There is nothing in the world of fine books quite like the first discovery of Audubon," later critic and connoisseur Sacheverell Sitwell concluded. "The giant energy of the man, and his power of achievement and accomplishment, give him something of the epical force of a Walt Whitman or a Herman Melville.... Audubon is the greatest of bird painters; he belongs to American history."[24]

Audubon finished with only about 160 subscribers, many fewer than he had hoped for. He estimated that the entire project had cost him in excess of $115,000, not counting living expenses for himself and family, over a twelve-year period. He probably received about $139,200 from his 160 subscribers (87 numbers x $10 x 160); he might have received another $4,500 from subscribers who did not complete the set, and he brought fifteen bound and several loose sets back with him, which he eventually sold for prices ranging from $874 to $1,100 each (depending upon whether they were loose sheets or bound in full leather). "*The Birds of America* being now positively finished," Audubon wrote his longtime friend Dr. Richard Harlan in 1839, "I find myself very little the better in point of recompense for the vast amount of expedition I have been at to accomplish the task."[25] Despite the fact that he did not sell as many subscriptions as he had hoped, sales receipts from *Birds of America* had enabled the Audubons to live well in both England and America, and they would provide him with an income as he began production of the long dreamed-of miniature edition.

CHAPTER THREE

The Royal Octavo Edition

History would have forgiven Audubon if he had retired to the simple life upon completion of the double elephant folio. He had given the world a masterpiece and could have found a peaceful niche in the forests of Louisiana, the Carolinas, or upstate New York, where he could have enjoyed his celebrity; Lucy could have returned to teaching, and the boys would have made their own ways. But Audubon never sat idle, and he turned his "holy zeal" on two other huge projects as soon as he got back to the United States. One became *The Viviparous Quadrupeds of North America*, envisioned as a publication on mammals on the same order as *Birds*; the other was the fulfillment of his "Great Work," the royal octavo edition of the birds.[1]

Audubon took up the miniature birds first, because it was a natural elaboration on the double elephant folio that would permit him to enhance that work as well as take advantage of the sales potential of the smaller work: he could illustrate all the species one to a plate and include the western birds that he had recently discovered. Little additional creative work would be required because the pictures and text could be taken almost entirely from the double elephant folio and the *Ornithological Biography*. Such a publication, done in the United States, as he explained to Dr. S. G. Morton, a friend and supporter in Philadelphia, would "secure for me the copyright, and put an end to all Spurious undertakings of that Nature in our Country."[2] The quadrupeds project, on the other hand, would require a great deal of additional research and painting, for he had only begun to gather the specimens. Nor was he as well acquainted with the mammals as he was with birds.

Ultimately, the most compelling reason for producing the octavo edition was the continuing need to develop other sources of income. Audubon had, of course, retained the copperplates for the double elephant folio so that no one else could use them, and he could reprint them if the opportunity arose. But, as Lucy explained to a cousin in

1839, it was unlikely that Audubon would be able to find an engraver in America who could print them, much less one who could match Havell's standards. Havell had, of course, immigrated to the United States and settled in New York City, but he was earning his living as a painter rather than a printer. The only assets that Audubon had were the fifteen bound sets of the double elephant folio, a few loose sets, and the $10,610.06 that twenty-nine American subscribers still owed him for numbers of the double elephant folio that they had received but not yet paid for. Whether he would be able to collect in such an unstable economic climate was, of course, conjecture.[3]

Immediately upon his return, he exhibited almost five hundred of his paintings at the New York Lyceum of Natural History, hoping that he would earn some gate fees and perhaps sell some of the folios. "These original drawings [*sic*] . . . are in all respects worthy the closest attention of the curious," the editor of the New York *Albion* ventured. "M. Audubon has devoted the greater part of his life to the study of the history of animated nature; and he has brought taste as well as talent to the subject." The exhibition, however, attracted only a few of the curious and resulted in no sales, leaving Audubon a "disappointed man." "If I had an extraordinary fat hog to pedestal, with a comfortable bed of straw, I could draw thousands from far and near," he later remarked to Samuel Breck, "but paintings, however beautiful or well done, will not attract enough people to cover the expense. In London I should be sure of constant visitors to my gallery, but not here." Those who might have declined to purchase *Birds*, thinking that Audubon might not be able to sell his remaining sets, or that it would be reissued and "that afterwards it would be cheaper," were wrong, Lucy wrote. "Already the mistake is beginning to be felt, since the coppers are all put by" and "the application of some for a few extra plates . . . cannot be [filled] . . . even now."[4]

Audubon eventually sold seventeen complete sets of the double elephant folio between 1839 and his death in 1851, but fearing that news of the miniature edition might harm the sales of the remaining folios, he initially revealed his plans only to the family and a few close friends. He was always concerned that other authors would swamp the market with their own bird books if they knew that he was planning one. "Swainson is publishing his incomprehensible Works," he had complained to Bachman in 1837. "Gould has just finished his Birds of Europe and now will go on with those of *Australia*. *Yarrell* [Plate 14] is publishing the *British birds* quarto size—and about one thousand other niny tiny Works are in progress to assist in the mass of confusion already scattered over the World."[5]

At home what some perceived as a threat came from Dr. John Kirk Townsend of

The Royal Octavo Edition

Philadelphia, a friend of several years who had accompanied Nuttall on Wyeth's second expedition to the Northwest in 1834 and had provided Audubon with several western specimens for the double elephant folio. Townsend's published narrative of his Oregon trip had sold out, and in 1838 he invited Audubon to coauthor an American ornithology with him. Audubon responded that Townsend should speak to Victor, who was already in America making preparations for the octavo edition, "as openly as if to myself" regarding his plans, leading some to suggest that Audubon was attempting to deceive a potential competitor, because he had no intention of collaborating with Townsend. When Audubon finally declined the offer, Townsend teamed up with lithographer J. B. Chevalier and printer E. G. Dorsey of Philadelphia to publish the first number of the proposed *Ornithology of the United States of America; or, Descriptions of the Birds Inhabiting the States and Territories of the Union* (Philadelphia, 1839). Apparently referring to Audubon's double elephant folio, Townsend wrote in his introduction: "The present publication is not expected to rival in their appropriate sphere those which have preceded it, but it is considered desirable to offer the public a work of portable dimensions and generally accessible form . . . with all the newly discovered species." Publisher Chevalier was more to the point, mentioning in a footnote on the back wrapper that the cost of Audubon's *Birds of America* was "upwards of $800!" Townsend proposed to issue numbers monthly until five volumes were complete, but the first number was so modestly produced, with only four hand-colored lithographic plates and descriptions, that the project failed from lack of subscriber interest. Copies are so rare today that a later ornithologist and bibliographer suggested that only a few copies might have been printed as a trial to see if it would sell.[6]

Had this publication been successful, it might have been a threat to Audubon's plans, and some have charged that indeed Audubon destroyed the market for Townsend's book by quickly issuing his own in similar covers. If that were true, Townsend showed no anger when he signed on as an agent for Audubon's book or wrote Victor in May 1840 to ask his assistance in getting the journal of his Oregon trip back into print. In the final analysis, there was probably little market for Townsend's book, because artist J. Delorme's prints were not in the same class as Audubon's (Plate 15). Audubon finally seems to have realized this and forgot about the entire endeavor.[7] He had been planning his "small edition" for years and saw no reason to give up the idea or share it with anyone.

Audubon's own set of the double elephant folio, now in the collection of the Stark Museum of Art in Orange, Texas, perhaps reveals some of the preparations that he had

49

The Royal Octavo Edition

made for the royal octavo edition. First, the set is not arranged as Audubon originally issued it but generally according to his concept of the appropriate scientific arrangement of the genera and species as represented in his *Synopsis of the Birds of North America* (Edinburgh, 1839), which serves as an index to the folio *Birds*. He and two close friends, Dr. Benjamin Phillips of London and Edward Harris, decided to bind their copies in this manner. He would use a similar arrangement in the octavo edition. Second, the set includes thirteen composite prints that Havell made in 1838 to comply with Audubon's request that he print those plates that needed "old or young birds . . . or females" added to them. In other words, Audubon was correcting errors or oversights, in which he had painted a species more than once or had not been able to secure a pair of specimens, by combining plates. These three sets would be closer to his initial desire to show the male, female, and young of each species in each plate. For example, plate 107 of the double elephant folio depicts the male and female grey jay (*Canada Jay*) (Plate 26). Plate 419 includes the young grey jay, which Audubon had drawn separately, thinking it was a different species (Plate 27). Plate 215 in Audubon's set is a composite print (Plate 25) in which Havell has combined these two plates so that all three jays are shown together. It is similar to the arrangement of the subsequent octavo plate (Plate 28), which contains all three birds, clearly copied from the composite plate. Audubon produced two octavo plates each from composites of grosbeaks (Stark 165, a combination of Havell 373 and 424 to produce Bowen 206 and 207), tanagers (Stark 168, a combination of Havell 354 and 400 to produce Bowen 209 and 210), and thrushes (Stark 193, a combination of Havell 369 and 433 to produce Bowen 139 and 143); three other octavo plates (Bowen 213, 214, and 218) were produced from the composite of blackbirds and orioles (Stark 172, a combination of Havell plates 388 and 433).[8]

Finally, there are pencil sketches and/or notations on four of the Stark plates—the *Louisiana Hawk* (Havell 392), the *Marsh Hawk* (Havell 356), *Indigo Bird* (Havell 74), and *Townsend's Sandpiper* (Havell 428)—but there is no clear relationship between them and changes that were later introduced into the octavo edition. The *Louisiana Hawk*, or Harris's hawk, has some sketching in the background, suggesting that Audubon might have thought of adding a landscape. To the *Marsh Hawk*, or northern harrier, someone added a rough sketch of a fourth bird with the notation: "Alter this hawk." The *Indigo Bird* has a notation at the lower right which reads, "This must be widened," and some sketching of unknown origin has been added to the *Townsend's Sandpiper*.[9]

Audubon planned the royal octavo edition along the general scheme of the double

The Royal Octavo Edition

elephant folio. He promised one hundred fascicules, or numbers, of five plates each, for a total of five hundred illustrations, sixty-five more than are in the double elephant folio. The hand-colored images were to be reproduced in the royal octavo size (one-eighth of a large sheet of paper), between 10¾ by 7 inches and 10 by 6½ inches, depending upon how the pages were trimmed for binding, and arranged according to genera and species. Audubon would incorporate texts describing the habits and localities of the birds, along with their anatomy and digestive organs (with occasional woodcut illustrations), taken largely from the *Ornithological Biography*. Each number would be wrapped in blue or grey paper covers (Plate 31) and distributed to subscribers for $1 each by Audubon and his network of agents on the first and fifteenth of each month. This new edition of *Birds of America*, he wrote, "will complete the Ornithology of our country, it is believed, in the most perfect manner." As with the double elephant folio, the "little work" was a family production, with thirty-year-old Victor managing all the family's business affairs from New York, twenty-seven-year-old John reducing the drawings with the camera lucida, and Audubon himself on the road selling subscriptions and contracting with agents.[10]

Once convinced that Havell would not undertake the project, Audubon looked elsewhere for a printer—and to another process. Copperplate etching had been the finest method of reproducing pictures for centuries, with aquatint used to produce halftones beginning in the 1770s. But by the 1830s, lithography, a less expensive and faster method of printing pictures, had virtually replaced etching and engraving, except for the finest and most expensive prints. Audubon had experimented with lithography while in Philadelphia in 1832, producing a print of the *Marsh Hens*, or clapper rail (19¾ by 26¼ inches), with Childs and Inman, and he had asked Richard Rathbone for the names of good lithographers as early as 1834. By 1839 lithography had improved to the point that the difference in the finished prints was apparent only to a trained and sensitive eye, and Audubon finally chose this method when he could not find an engraver in America who could match Havell's skill.[11]

It might have been to assist those involved with Townsend's failed effort that Audubon inquired of Chevalier and Townsend whether they intended to continue their publication. Since Audubon expected to spend most of his time on the road selling subscriptions, he would need someone to help Victor manage the project. Chevalier had good contacts in Philadelphia, particularly with the press, and had even officed at 72 Dock Street, also the home of the influential *Saturday Courier*. Audubon arranged with him to act as his agent or business manager. His name would appear on the title page as copublisher along with Audubon. Townsend became one of the sales agents for the new work. Audubon contracted with E. G. Dorsey, who was also the printer for Townsend's book, to print the text for *"la petite Edition."* Then he chose John T. Bowen, also of Philadelphia, to reproduce the birds by hand-colored lithography.[12]

Bowen, along with George Endicott of New York, Peter S. Duval of Philadelphia, and John H. Bufford of Boston, was one of the most outstanding lithographers in America. An English immigrant who was born in 1801, he was something of an artist in his own right and was working as a print colorer in New York as early as 1834. He expanded into lithography and moved to 94 Walnut Street in Philadelphia in 1838 to take over production of Thomas McKenney's and James Hall's *History of the Indian Tribes of North America*, a series of 120 brilliantly colored portraits copied after paintings by Henry Inman and Charles Bird King. It was the most ambitious lithography project in the United States at the time, and he employed the best printmakers and colorers in the city for the task. He had successfully completed the stalled first volume, which McKenney had begun with Childs and Inman in 1830 and continued with Lehman and Duval, and was at work on the remaining volumes when Audubon chose him, presumably at Chevalier's recommendation and probably because he—Audubon—had seen the Indian portraits, to print and color the illustrations for his octavo *Birds*. Bowen was at the height of his career and although the records of his business are not known to exist and we cannot be sure, he no doubt employed scores of workers. Two years later, when Philadelphia artist Joseph Sills visited the Bowen shop, he found it "altogether very interesting, from the Press Room below to the sizing Room, colouring Room, and Lithographic Drawing Room."[13]

Lithography reached America shortly after its invention by Bavarian Alois Senefelder in 1798. Bass Otis, who is generally credited with the first American lithograph, a tiny and naive view of a mill that appeared in the *Analectic Magazine* in 1819, probably imported the entire process directly from Senefelder, who had published an English translation of his pioneering work on the subject that year. Two years later Barnet and Doolittle in New York produced lithographs for the *American Journal of Science and Arts*, and the following year, they printed twenty-one rather humble hand-colored plates for what probably was the first American book illustrated with lithographs, Sir James Edward Smith's *The Grammar of Botany* (New York, 1822). The industry grew rapidly from that timid beginning and had surely, by 1839, replaced engraving and etching, which never reached the heights in the United States that they had achieved in Britain, as the best method of reproducing pictures.[14]

The Royal Octavo Edition

Lithography is often explained in simple terms, but it is a deceptively complex process that eluded many would-be printers.[15] It requires a soft, porous stone, the best of which is still quarried near Solenhofen, Bavaria. The stones are cut into various sizes and thicknesses, depending upon the size of the press and the image to be reproduced, and then ground (one stone against another, with increasingly fine layers of sand mixed with a few drops of water in between) until the desired texture, or grain, is obtained. The lithographic artist, who does not need the special skills of the engraver, then draws on the stone with special greasy or waxy crayons. When the drawing is complete, the stone is bathed with gum arabic and nitric acid to "fix" the drawing. The stone absorbs water, except for those parts covered with the greasy—and, therefore, water-resistant—drawing. A greasy, sticky ink is then applied to the stone with a roller. It adheres to the drawing but is repelled by the wet stone. The printer then lays a piece of paper on the stone, applies pressure, and "pulls" the print from the stone (Plate 30).

Pressure may be applied by hand (rubbing a block of hard, sanded wood across the entire surface of the print) or with a press. Lithographs could not be printed in letterpresses or copperplate presses because far more pressure was needed for a lithograph than for ordinary letterpress. If sufficient pressure to print a lithograph was added to the letterpress, meaning that pressure would be applied to the whole stone simultaneously, the stone would crumble. The rollers of the copperplate press, which worked fine with a thin metal plate, tore the paper when a thick stone was used. The most popular press in America during the 1840s probably was one in which the stone was placed in the bed of the press and passed beneath a fixed scraper blade, which applied immense pressure to the print and the stone, but only where they were touched by the blade.

The great achievement of lithography was that the artist could draw directly on the stone and it would mirror his or her strokes exactly. No intervening engraver was needed—whose limitations sometimes significantly altered the artist's style, as in the case of Audubon and Lawson. The resulting reproductions, usually in black ink, are so accurate that artists began to refer to them as multiple originals rather than copies or reproductions. Lithographers, of course, have developed many refinements until literally thousands of prints may be pulled from a single stone, depending upon the quality of the stone, the drawing, and the ability of the printer.[16]

The unsung heroes of American lithography are the pressmen, artists, and colorists who produced the literally millions of hand-colored illustrations that supplied a growing public with beautiful images. In preparing the reductions of Audubon's images, they faced a daunting task: to copy, in a greatly reduced format, one of the most beautiful books ever produced. One of the most difficult problems of the industry prior to the development of photography and light-sensitive printing plates was enlargement and/or reduction of an image and copying it onto the lithographic stone. Sometimes lithographic artists employed Audubon's method of grid lines to help them transfer a drawing to the stone. Many, like John Woodhouse Audubon, used the camera lucida for reduction, but would-be geniuses toyed with all kinds of possibilities, including stretching a rubber sheet inside a frame by means of evenly spaced, adjustable hooks. The image to be reproduced was transferred to the rubber sheet, which would then be stretched or reduced by adjusting the hooks. The resulting image would then be transferred from the sheet to the stone.[17]

John was assigned the task of reducing the life-size Havell prints by the more common method of the camera lucida. Designed by English scientist William Hyde Wollaston in 1807, the camera lucida (Plate 29) was nothing more than a carefully positioned glass prism held at eye level by a brass rod over a flat piece of drawing paper. Looking through a peephole centered over the edge of the prism, the artist could see both the object to be drawn and his paper. John's drawing, therefore, was a virtual tracing of Audubon's original painting, except that it was much reduced in size, the reduction being determined by the distance between the subject and the prism and the prism and the drawing paper.[18]

This resulted in an outline drawing that Bowen's artists traced over with a red or brown pencil, then placed face down on the stone and rubbed the back, transferring or offsetting the drawing to the stone. Next, the lithographic artist would fill in the details to reinforce and complete the drawing. In an alternate procedure, he might seal a piece of tracing paper over the drawing with a large dollop of wax or glue in each corner and trace the drawing. The tracing paper would then be placed, face down, on the stone and rubbed to offset it onto the stone. Bowen usually used Havell's prints as the pattern, but he used the original watercolors on occasion. Once, for example, Victor concluded that they were copying the aquatint effect that Havell had introduced into the print and suggested that Audubon provide Bowen the original watercolor so his artists could imitate Audubon's more fluid strokes. A number of the extant drawings made for the octavo edition have a grid pattern in the background, reminiscent of the technique that Audubon developed as a young man, but in this case more characteristic of an artist attempting to overcome the distortions inherent in the camera lucida. Thirty-two drawings used in the production of the octavo edition are

The Royal Octavo Edition

now available for study in the collections of the Hill Memorial Library at Louisiana State University. Many of them have been traced over in brown pencil or chalk, and almost all of them show rub marks on the back, probably the result of transferring them to the stone. Many also have lumps of glue and wax remaining on the face, and a couple still have bits of the tracing paper embedded in the wax.[19]

Beginning cautiously, as he had with the double elephant folio, Audubon ordered only three hundred copies of the hand-colored lithographs for the first four numbers in November 1839. Bowen charged $7 each for R. Trembley, and later W. E. Hitchcock and other artists, to copy or transfer John's small drawings onto stone and $27 per one hundred for the paper, printing, and coloring, for a total of $116 per number, or almost $.08 per print. Bowen and Audubon agreed that fifty percent would be paid upon delivery of the prints and the remainder within thirty days. Audubon ordered twelve hundred copies of the much less expensive letterpress. Dorsey charged $30.27 for printing twelve hundred copies, with paper for text and covers costing an additional $129.79. Edward H. Rau charged $6.25 for the tissue paper to cover each plate and to bind, trim, and pack each number. Audubon's total cost of production, including the $7 per number that he paid John for making the small drawings, was approximately $.57 per number, the bulk of the cost going for the lithography (approximately $.39 per number) and paper (approximately $.11 per number).[20] These prices were comparable with those charged for similar projects by a number of different lithographers.[21]

Bowen printed Audubon's small birds two to a stone in black ink. He delivered 300 copies of numbers 1, 2, and 4 with number 3 arriving in a first printing of 210, for a total of 5,550 prints (300 x 3 + 210 x 5) early in December. (See Table 1.) Even with these modest orders, Bowen soon found himself with tens of thousands of prints to color. He turned them over to teams of colorists, usually young women, who probably worked similarly to those in the Currier and Ives shop in New York. They sat at long tables with a model, usually a colored copy of the print that Audubon or Victor had approved, sometimes one of Havell's prints or Audubon's original watercolors, in the middle of the table for all to see. Each colorist would be assigned a section of the image and would apply only one color. The print would then be passed to the next person, who would apply a different color. Finally, the "finisher" did what was necessary to complete the picture, usually only touch-ups and highlights. If large numbers of prints were needed, Currier and Ives would sometimes cut stencils for the various colors, which could be washed in by unskilled help outside the shop and then finished by one of the regular employees. Currier and Ives paid $.01 each for coloring the small prints, $1 for twelve of the large folio prints.[22]

For large jobs such as Audubon's, Bowen employed similar help in Philadelphia. Many of the young women would take a colored octavo proof home, where they worked. More than two hundred of these proofs, now in the collection of the Stark Museum of Art, reveal this aspect of the coloring process more fully. The correction process is evident on the plate of the merlin (Bowen 21), for example, where someone, perhaps one of the Audubons, perhaps Bowen's chief colorist, has written "Not so much blue, more grey," on the back of the bottom bird. On the American coot (Plate 23) someone has written: "Use green over the *head* in the background…deeper & bluer to define the outline of the bill." "Spare the blue on leaves and do not keep them any stronger," another wrote on the proof of the least flycatcher (Plate 24). Then, in a blue pencil, another hand has added, "Do the leaves as near as you can & omit the blue shade." A number of the plates are marked "Pattern" or "Good Pattern." There was much back and forth between Audubon and Bowen and between Bowen and his employees.

These proofs also identify some of the anonymous colorers who produced such handsome work for Bowen and Audubon. Some bear the initials "A.V." (Bowen 265 and 268). Others are initialed or signed by Mary Tallon (Bowen 275, 314, 335, 336, 365, 385, 394, 399, and 404, which also contains her address: 1132 S. 11th Street, Philadelphia) and Amanda E. A. Sherwell (Bowen 353 and 388).[23]

The results justified the effort. The Reverend Bachman pronounced himself "delighted" with the new prints in November 1839, probably after seeing the first proofs. Audubon sought a further comment in December: Should he increase the amount of text? Should he add the new birds that he had discovered in the order of the *Synopsis* or put them in an appendix? Hearing some comments that he apparently considered a bit too critical, Audubon responded sharply the following month:

> It appears from what you say that you expected that I should have given (in the letter press) the whole of my Knowledge in the Ornithology of our Country; when in fact, this never entered my head. No! I have further prospects in view, and for the present, I thought (and that after rather considerable reflection) that to abridge the amount of my former letter press, to meet the size and general appearance of the present publication, none or few of our subscribers would complain (especially as very few of them are Naturalists) but on the contrary would

be pleased, that in my so doing I saved them a considerable addition of expense. . . . My Intentions are to give the best of figures of our birds, taken from my own Original Drawings, and to give figures of all and every Species which are *Now Known* or may *become Known*, between the present day and the last day of Said publication.²⁴

But Bachman was not subdued: "The descriptions in the 'Small Edition of Birds' will have to be abridged—your '*worthy friend*' and other humbugs may be left out to advantage."

Then he offered Audubon some advice on the subject of subscribers:

Cities are not the only places to obtain them. Birds sing and nestle among the groves of the *country*—The planters and farmers are the men to become subscribers. An intelligent planter from the up country said, a few days ago, that if the right person would thoroughly canvass the whole State of South Carolina, he would insure three hundred subscribers to the "Small Work." *Old Jostle* [Audubon] would be the man, and when his legs failed, the *Young Jostle* [his sons] should go forward. Get the Editors to notice your work—this is a puffing world—from the porpoise to the steamboat.²⁵

Bachman apparently set out to provide some of the puff himself in a review of the first number that he probably wrote that same month for the *Southern Cabinet* in Charleston.

Audubon, whose inimitable drawings and accurate descriptions, have brought . . . [ornithology] to a high state of perfection[,] . . . has now commenced publishing his great work on American ornithology, in a reduced size, and according to a scientific arrangement, giving good figures on stone, and all the information contained in the larger work. . . . At one dollar per number, [it] . . . is decidedly the cheapest work on material history ever published in any country. . . . The work cannot be superseded by any other, and it will remain a standard work for ages to come.

He continued the following month: "The second number . . . is a beautiful specimen of lithographic engraving, which is now brought to great perfection. . . . The coloring of this number is superior to the first, and as the publication progresses, every number will unfold new beauties.... Any one who is able to read our language may become an ornithologist."[26]

The octavo edition was, indeed, impressive. It was the most comprehensive accumulation of research available on American birds. It appealed to natural history enthusiasts as well as anyone who loved America. As the New York *Albion* explained, it was "fitted both in size and price for general circulation." And it was beautiful. Audubon had indicated his awareness of the aesthetic qualities of his work much earlier when he commented on the impact of Havell's engravings. He attempted to maintain that same look in the miniature edition by simplifying the prints and elegantly reproducing them on a large, octavo page, backed, with few exceptions, with pristine white. One critic has suggested that the intensified color that results is similar to the enhanced black-and-white effect that fashion photographer Richard Avedon achieves today when he poses his portrait subjects against a white, seamless cloth. Audubon also lavished attention on the text design, providing enough margin and space between the lines that the overall effect is pleasant and easy to read—at a time when a physically beautiful book would have been viewed as an art object in itself. He chose a quality paper and a typeface that he considered "new, clear and I think very fine" for the *Ornithological Biography* and no doubt felt that the typeface selected for the octavo edition was equally as handsome. He also paid close attention to the design of the page and the "arrangement of the formulas, Headings &c."[27]

Using the same process he had worked out for the double elephant folio, Audubon began the search for subscribers to the "little work" as soon as he had samples to show. He was a master salesman, having honed his skills on the much more expensive book over a period of ten years, and as Bachman had suggested, it seems that he did everything right. This time he began with substantial name recognition and a reputation, and he was selling a much less expensive book. While his celebrity status may make him too much of an exception, his methods may be examined as a representative case study in the selling of subscription natural history books in pre–Civil War America. He continued to use his memberships in learned societies and strategic letters of recommendation to convince the naturalists of his qualifications and to meet people who would be predisposed to want his book as well as those who could afford it. Letters from three individuals, he said, led to his great success in Baltimore. The fact that the octavo edition cost about one-tenth the price of the double elephant folio made a big difference as well. Even if a subscriber could not afford the double elephant folio, he or she could still participate in Audubon's adventure and own a beautiful object at the same time.[28]

The Royal Octavo Edition

While Audubon was on the road, he made arrangements with agents to represent his publications on commission. He had several contracted when he issued his prospectus and signed up others as he went, generally offering a ten to twenty percent commission.[29]

Perhaps the most important element in Audubon's success was his own personality. Describing him shortly after his return to America, an unsigned author, probably Bachman, writing in the *Southern Cabinet*, commented that "although the snows of age are gathering on his head, and here and there a furrow may be seen on his cheek, yet his eye has still the brightness of a hawk, he treads with the firm step of a young man, and his energies are unimpaired." Samuel Breck of Philadelphia, who met Audubon at the Academy of Natural Sciences in Philadelphia late in 1839, estimated that he was "a man of fifty." Actually he was a hearty fifty-four, the years of hunting and strenuous living giving him a ruddy complexion and athletic build that belied his age. He had "the countenance of a bird," Breck continued, with "a projecting forehead, a sunken black eye, a parrot nose, and long protruding chin, combined with an expression bold and eagle-like." He was the sort of guest that hostesses sought to ensure the success of their dinner parties and as a result, he rarely dined alone while on the road. Having overcome his earlier shyness, he was a raconteur of legendary abilities at the dinner table and a gadabout at his exhibitions. Parke Godwin described him in 1842 as "noble and commanding" with an expression that "made you think of the imperial eagle." The Philadelphia *Mercury* noted in 1843, his fifty-eighth year, that he "attracted general observation" on the streets and that he reigned as "the lion of the afternoon" at a popular pub. Audubon had enough of an ego to realize that he was a pivotal figure. "It is true that my Name, and as Victor is pleased to say my 'Looks' May have some Influence in the Matter," he wrote in 1840.[30]

While Audubon had his detractors, many of whom published criticisms of his work in natural history journals, surprisingly few of their comments reached the popular press. Among the exceptions are John Neal's articles that appeared in the *New England Galaxy* and *Brother Jonathan* during the mid-1830s. Neal was a subscriber to the double elephant folio who became dissatisfied with Audubon over price, bindings, and what he considered to be unfulfilled promises. In the meantime, he had met and corresponded with Joseph Mason, who was upset because Audubon had not given him proper credit for his work. Mason assured Neal that Audubon had said that Santo Domingo, not Louisiana, was his birthplace. Neal ridiculed Audubon's stories of the wilderness hardships that he had suffered on behalf of his dream and doubted that he

The Royal Octavo Edition

had ever met Daniel Boone. What if the work turns out to be no more reliable than its author, he asked. Although Neal admitted that "Mr Audubon has published the most magnificent work in the world," he discontinued his subscription because he had "lost all confidence in the Author." Judge James Hall also raised questions in the *Western Monthly Magazine* about some of Audubon's published stories.[31] Their comments were all but forgotten by the time the octavo edition began to appear.

Audubon also knew the importance of well-placed, friendly articles in the press. He had used the procedure successfully in England and began to cultivate influential editors as soon as he returned to this country. He regularly gave review copies of the publication to the Philadelphia *Saturday Courier* and the New York *Albion* and frequently visited the New York *Evening Post* to talk with editor William Cullen Bryant. The New York *Albion* responded, noting on January 25, 1840, that:

> *Mr. Audubon* has resolved upon issuing an edition, reduced in size, but complete and carefully coloured, of his great national work, "The Birds of America." The splendid folio edition, although highly important for the full development of the subject, is evidently too extensive for general attainment; yet the subject itself is of too prominent interest to remain in the comparative obscurity incidental to a splendid and high-priced edition. The present therefore is issued in octavo. . . . We trust the opportunity will be widely seized, to acquire so valuable an accession in American Natural History.

He also published the prospectus for the book in several publications.[32]

Audubon and his friends were always on the lookout for influential regional papers, such as J. D. Legaré's *Southern Cabinet* in Charleston, and knew how to take advantage of one story to encourage others. Mrs. Audubon wrote of a positive story that had appeared in the Baltimore *Patriot* in early 1840 that "Victor makes all the use of it he can." And Victor indicated, in a postscript, the contacts that he had made: "Col. Stone has promised to notice the work. So has Dr. Bartlett [of the *Albion*] & Mr. McDaniels." Audubon reported that the article in Legaré's journal had been copied "*in all* the papers here [and] has produced a capital effect. It is . . . the best notice that has appeared."[33]

Audubon realized he had a success on his hands after his first sales trip to New England. "Since my arrival here [I] have been in constant Bustle, moving as fast as my legs could carry me, ringing at many many doors, and shaking the hands of Numberless Worthy Men and fair Ladies!" In less than three weeks he had sold ninety-six

22. R. Trembly after John James Audubon, *Fork-tailed Flycatcher*, 1840. Hand-colored lithograph by Bowen, 10 1/4 by 6 1/2 in. (sheet). From *The Birds of America*, plate 52. Courtesy W. Thomas Taylor, Austin, Texas. Audubon credited Maria Martin with the beautiful painting of the loblolly-bay.

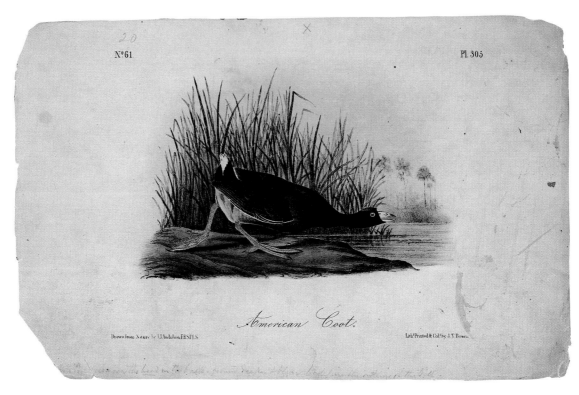

23. J. C. after John James Audubon, *American Coot*, 1842. Hand-colored lithograph proof by Bowen, 7 by 11 1/16 in. (sheet). From *The Birds of America*, plate 305. Courtesy Stark Museum of Art, Orange, Texas. The instructions written on the proofs of the octavo plates show many of the corrections necessary for Audubon to get the proper coloring on the birds. Here someone has written, "Use green over the *head* in the background...deeper & bluer to define the outline of the bill."

24. R. Trembly after John James Audubon, *Least Flycatcher*, 1844. Hand-colored lithograph proof by Bowen, 10 11/16 by 6 15/16 in. (sheet). From *The Birds of America*, plate 491. Courtesy Stark Museum of Art, Orange, Texas. On this plate someone has written, "Spare the blue on leaves and do not keep them any stronger." Someone else added, "Do the leaves as near as you can & omit the blue shade."

25. Robert Havell, Jr., after John James Audubon, *Canada Jay* (composite plate), 1838. Hand-colored aquatint and engraving. From *The Birds of America*, plate 215 in the Stark Museum set. Courtesy Stark Museum of Art, Orange, Texas.

26. Robert Havell, Jr., after John James Audubon, *Canada Jay*, 1831. Hand-colored aquatint and engraving, 26 by 20 3/4 in. (plate). From *The Birds of America*, plate 107. Courtesy Stark Museum of Art, Orange, Texas.

27. Robert Havell, Jr., after John James Audubon, *Canada Jay*, 1838. Hand-colored aquatint and engraving, 19 1/2 by 12 3/8 in. (plate). From *The Birds of America*, plate 419. Courtesy Stark Museum of Art, Orange, Texas.

28. R. Trembly after John James Audubon, *Canada Jay*, 1841. Hand-colored lithograph by Bowen, 10 1/4 by 6 1/2 in. (sheet). From *The Birds of America*, plate 234. Courtesy Stark Museum of Art, Orange, Texas. Audubon originally depicted two grey jays in plate 107 and an immature one in plate 419. He added the immature bird from plate 419 to the pair in plate 107 to produce the composite plate (plate 25), which he then used as a guide for the octavo plate.

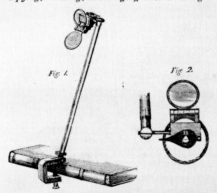

29. *The Camera Lucida*, undated. Broadside. Courtesy Gernshim Collection, Harry Ransom Humanities Research Center, The University of Texas at Austin. Audubon acquired and began using a camera lucida soon after his arrival in Britain. Broadsides such as this one explained how to use it.

30. Louis Prang, *Lithographer*, 1874. Chromolithograph, 14 by 21 11/16 in. Courtesy Amon Carter Museum, Fort Worth, Texas. Boston lithographer Louis Prang illustrates four steps in the lithographic process (left to right): grinding the stone smooth, drawing the image on the stone, preparing the stone for printing, and printing.

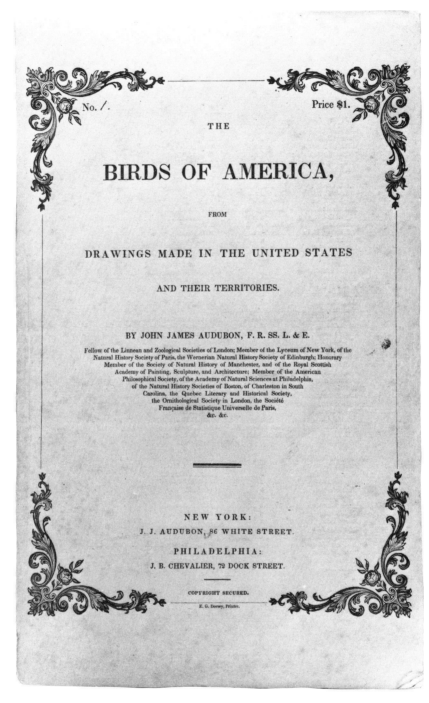

31. Cover of Audubon's royal octavo edition of *The Birds of America*, 1839. 10 3/4 by 6 7/8 in. Courtesy Stark Museum of Art, Orange, Texas. Audubon paid close attention to the overall design of his book and secured an attractive format for the cover. It was, as some critics charged, similar to the cover that Townsend had used on his short-lived project.

subscriptions in Salem, Boston, and New Bedford. "We did indeed always *anticipate* it, but there is nothing like the *reality*," Victor resolved in December 1839. Even the binder's spoilage of the entire printing of 300 copies of the second number did not dampen his enthusiasm. "This mischief will be got over next week," he informed Bachman, who was also helping with the selling, "and then we will forward you the usual quantities." To the Reverend Bachman, Audubon wrote, "Tell our good Sweet heart [Maria Martin], that I am much in want of drawings of plants (such as I have not published) of any kind and drawn on paper the size of *the Small plates of our Large Work*." The number of subscribers had climbed to 300 by the time he wrote Bachman again, and he had ordered additional copies of the first four numbers. The increased print order of 500 copies for number 5 arrived on February 20, 1840. (See Table 1.) "This beautiful work and its very superior advantages we have already fully described," the *Albion* reported on February 8. "The newspapers here have began [*sic*] a good firing and I expect to do something next week," Audubon wrote the family on February 15. "The publication of the little Work" is "going on quite well," he told Victor in March. "The proofs of No 7 are quite beautiful." Audubon increased the print order to 750 (3,750 prints) with number 8; 1,000 (5,000 prints) with number 10; and 1,250 (6,250 prints) with number 24.[34]

Bowen reduced the unit price slightly as he increased the print run. He charged $.34 per number for 500 copies of each print, $.32 for 750 copies, $.31 for 1,000 copies, and $.304 per print for 1,250 copies. Audubon chose a better quality of paper beginning with number 9. He also ordered more letterpress. Dorsey had printed 1,200 copies of number 1, 1,000 copies of number 2, and 500 copies of numbers 3 and 4. It appears that Audubon maintained that number through number 8, despite the fact that Bowen was providing 750 copies of the lithographs beginning with number 7. Apparently he returned to 1,000 copies of the letterpress beginning with number 9, and in December 1840, Victor urged that he increase it to 1,100 at number 20 or 21 and subsequently to 2,000.[35]

As Audubon headed south, the reaction in Baltimore was even more than he expected. He obtained 171 subscriptions there alone—"within thirty days we have doubled the Number of our subscribers, and now have above *Five hundred and Twenty*!" —and was soon urging Bowen to increase the pace of production. "The amount of attention which I have received here is quite bewildering, the very streets resound with my name, and I feel quite alarmed and queer as I trudge along," he confessed to the family on February 21. "But alas! I am now out of numbers to deliver to my subscribers here," and he had not even called on all the persons on his list. "That one single City of moderate Comparative population should have supplied me with a list of Subscribers to a Work amounting to one hundred Dollars to the Number of One hundred and fifty Seven in the Course of one fortnight, is, and probably will ever be unprecedented."[36] He eventually got a total of 168 subscriptions in Baltimore. (See Table 2.)

His problems were now those that come with success. To fill these orders, and the additional ones that he expected, for he had yet to visit several parts of the country, Audubon suggested that they suspend publication of the new numbers for one month so Bowen could reprint the first four numbers. Victor disagreed: "I am sorry you suffer yourself to be annoyed about the [back] numbers for the new subscribers," he wrote in March. "They *must wait* and *I advise you as you procure new names, to say so at once*—a week, cannot signify much to them, but we will damage our publication by interrupting the regular course of it." Victor then informed Bachman, "We are put back a little by the rapid increase of our subscribers, but now I think will be able soon to meet every demand as Bowen promises me to devote his whole attention to it."[37]

The public continued to be enthusiastic. Eight subscribed at Annapolis, fifty-three in Washington, and twenty-nine in Richmond. Audubon quickly signed up five subscribers upon his arrival in Charleston and authorized a third agent to operate in that city, hoping to take advantage of the fact that the man was also the agent for a popular magazine. The country that had initially shunned his "Great Work" (and eventually purchased only eighty-two copies of it) now ordered hundreds of copies of a publication that would eventually run to seven volumes and that, at the time, retailed for $1 per number for one hundred numbers, unbound, faster than he could produce them.[38]

Audubon's journal provides an entertaining account of his travels as well as insights into his success. By the late summer of 1840, he was back in New England. He went to New Bedford, Plymouth, Duxbury and Hingham, Boston, Lowell, and Portsmouth before returning to Boston, where he remained until he learned of the death of John's wife, Maria, on September 15 in Charleston. John had taken her there, hoping that the warmer climate would help her recover from tuberculosis. Spending October at home in New York, Audubon was back on the road by the first week of November. Accompanied by Havell, he first went to Philadelphia to check on progress of the book, dining that evening with Bowen, Chevalier, Havell, and the editor of the *Saturday Courier*. Passing through Baltimore and Washington before returning to New York, he was able to check with several of his agents and visit with Colonel John James Abert, head of the topographical engineers, and several other friends in government. Audubon

The Royal Octavo Edition

probably was still hoping for a place on a government expedition into the West.[39]

His next tour of New England pushed the orders to almost a thousand, convincing him that he might ultimately be able to secure as many as two thousand subscribers. He stopped briefly in New York to pick up additional copies before sailing up the coast to Providence and Boston on November 16, where he delivered the new numbers and settled accounts with agents such as the young firm of Little and Brown, which specialized in importing European and British publications. In Boston Audubon called on Senator Daniel Webster, who was notorious for the disarray of his financial affairs. He reluctantly paid $100 on the set of the double elephant folio that he had purchased in 1836, but nevertheless, he subscribed to the octavo edition, promising to pay for it the following week. He also assured Audubon of additional letters of introduction to friends in Worcester, Springfield, Hartford, New Haven, Northampton, and Albany. "*Nou verrons!*" Audubon wrote in his journal. Audubon remained in Boston until December 12, enjoying the social and intellectual life of New England's most stimulating city and calling on would-be subscribers, agents, and booksellers.[40]

He paused over the weekend in Worcester, visiting the insane asylum and the "curious old Books, paintings, etc. etc." in the Antiquarian Society. Since no train departed until Monday, the society librarian urged him to visit with Elihu Burritt, the "famous learned Black Smith," a then well-known advocate of world peace. Audubon found Burritt "with his sleeves turned up, his arms bare and full of sinew, his eyes sparkling bright, his forehead smooth and high, his person manly, his demeanor modesty itself! We shook hands," Audubon continued, "and talked for a while, and strange as it may appear, his opinions of our success in regard to mental improvement coincide precisely with my own, i.e., *We are what we make* ourselves. He asked for my autograph and I wrote a few Lines being a very poor exchange for his Signature on my Subscription List." The New England winter seemed to relent a bit as Audubon moved on to Springfield, Hartford, and New Haven before returning to New York to be with the family through the remainder of the winter. In little more than a year, Audubon had sold almost nine hundred subscriptions himself.[41]

After recalling that he had reviewed approvingly the first volume of the series, an anonymous critic for the *American Journal of Science and Arts* in New Haven commented on volume two:

> To praise it is no longer necessary; for . . . the public have given indubitable assurance that his labors have been appreciated, . . . and . . . it can hardly be said of

him that he "is not without honor save in his own land." . . . Mr. Audubon has now nearly a thousand subscribers to his work, an instance of liberal support of a work on natural history certainly without parallel in the New World, and hardly with one in the Old. This insures the success of the undertaking far beyond the most sanguine anticipation of the author, and enables him to continue to marked and decided improvements in the publication as it advances. . . . We witness no abatement in the interest or the value of the work.

Audubon began to refer to the new work as his "Salvator."[42]

CHAPTER FOUR

A Great National Work

Even though the octavo edition was based extensively on the double elephant folio and the *Ornithological Biography*, overseeing and producing it still was a complex job that demanded organizational as well as administrative abilities. Chevalier oversaw production in Philadelphia, and Victor managed the business and family affairs from New York, but Audubon had the final word in all matters related to his work.

By the time the fifth number appeared, problems began to arise. The craftsmen who lettered the plates had not followed Audubon's instructions precisely and had confused three bird names. Victor instructed them to change "Pigeon *falcon*" to "Pigeon Hawk," "*Sparrow falcon*" to "Sparrow Hawk," and "*Gos Hawk*" to "Goshawk," and to abandon the block lettering they were using in favor of a heavier and much handsomer script. Unfortunately, the first printing had been distributed with the errors (and curiously Audubon seems to have changed "Sparrow Hawk" back to "Sparrow Falcon" and "Goshawk" back to "Gos Hawk" in later issues).[1]

What could have been a devastating problem never materialized. Audubon had worried about getting his *Birds* copyrighted in the United States ever since he began the project and had printed volume one of the *Ornithological Biography* in Philadelphia soon after it appeared in Edinburgh to gain the protection of an American copyright. One of his reasons for rushing the octavo edition into print was to get the entire work under an American copyright. Then in March 1840, according to Henry Havell, Robert, Jr.'s brother, it now appeared that "some rascal" might beat Audubon to the task. Someone had apparently requested a bid from Henry for coloring copies of Audubon's bird prints, which they intended to sell for $.37½ per number, only a few pennies more than Audubon himself was paying! Worried, Victor urged his father to ask Congress to pass "a bill granting the *exclusive right* [to *Birds of America*] to *us* . . . for as long as a term of years as you think can be obtained." But Audubon scoffed at the idea—

the "story about another edition of our Work is perfect Nonsense. Where would the Rascal procure Subscribers . . . ? Do not be Startled at a Scare Crow!"—and when Havell reported that he had not heard from the person again, Audubon told Victor to drop the matter.[2]

More serious was Bowen's threatened bankruptcy, which probably gave Audubon flashbacks of the colorers' strike that had forced Lizars to give up production of the double elephant folio. Just at the time Audubon needed the printer to devote all his attention to production, creditors were calling on him for payment of his bills and threatening to force him out of business. The McKenney and Hall Indian portfolio, Bowen's other big job, was the problem. The publishers had gone bankrupt, and Bowen had not been paid. He tried to publish it himself, so he could recoup some of his expenses but was forced to stop. Audubon and Victor spent a little time searching for another lithographer who might at least supplement Bowen's production, if not replace him entirely should he fail. Victor reported that Henry Havell could color two hundred to three hundred of each number in New York if needed, but that when Bowen had approached him about the job, Havell had replied that he would work only for the Audubons. "I am sorry, very sorry that that foul Hy. Havell should have refused to Work for *us* through Bowen," Audubon replied, apparently failing to understand why Havell would not want to work for someone on the verge of bankruptcy. Later that month, Audubon, Chevalier, and a friend at the *Saturday Courier* loaned Bowen money to help him out of his predicament, and Audubon concluded that his "Situation after all does not appear to me to be as desperate as we would have supposed from the tone of his letter." Audubon finally concluded that "Bowen cannot . . . go through the Work as fast as we wish for the want of Colourers."[3]

The last problem was even more difficult. The Audubons had long known of the history of tuberculosis in the Bachman family, and Maria and Eliza were frail when John and Victor married them, but no one expected that it would strike so soon. Just as the success of the octavo edition was becoming apparent early in 1840, Maria's health began to worsen, and John took her back to Charleston, hoping that the warmer climate would help her recover. The family feared that John's production of the small drawings would suffer as a result, and their concerns were soon confirmed: Chevalier soon reported that some of the drawings that John had sent were so poorly composed that Bowen's artists had to redraw them, prompting Victor to write John that he should be "careful, as the camera undoubtedly produces distortions & differences in size of *each different* bird on the same plate." Bowen's artists probably had to redraw

A Great National Work

the small outline drawings that the Audubons sent them on several occasions, for two of the extant drawings bear the signatures of other artists: the *Mourning Ground Warbler* (Bowen 101), signed by R. Trembley, who did more of the Audubon plates than anyone else, and the *Black-Throated Wax Wing* (Plate 33), which bears the signature of an artist who signed himself or herself "W."[4]

John's production soon ceased altogether, even though the family continued the none-too-subtle suggestions that he keep up the work. Audubon wrote on March 9 that he had gotten a total of 171 subscriptions in Baltimore and 8 in Annapolis and that he was headed south in anticipation of gathering even more. On April 12 he included the news of fifty-three more subscribers from Washington. Victor added to John's burden on April 23, reporting that Havell wanted payment of the $1,000 owed him and bluntly warned, "You see my dear John, therefore the necessity of energy, as it *will not do for us* to let the *grass grow under our feet!*"[5]

Devastated by Maria's illness, John was unresponsive. In June Victor warned that "Papa says you had better begin to draw again as soon as you can. I wish you were likely to be able to come to us before July is out, for I think every day you are away from home now, such a sacrifice." A month later he admonished: "We shall want small drawings again in the month of October at latest," and he continued the barrage the following week, reporting that Audubon was in Boston, "where he is getting 8 or 10 subscriptions a day!" As if that were not enough, Audubon prodded Victor at the end of July to "write to Johny to draw small Drawings. I wrote a long letter to him on Sunday from Nantucket."[6]

In August, John began to work again. He sent a painting of a deer for the quadrupeds project and his portrait of the Reverend Bachman for the family to examine and critique. He also mentioned that he had made a drawing of the "new Oriole" from the West Indies or Mexico. Victor responded that, "You can do as many small drawings as will occupy you, and I think you would find it a relief to have none to do for some time after your return home." He also offered some encouragement from Chevalier, who had written that, "Whoever made these last small drawings, has my compl[i]ments, they are infinitely better drawn than those received previously, which bye the bye we had to redraw [e]ntirely for want of accuracy and proportion." Knowing that Maria's health continued to deteriorate, however, John answered despondently on September 2 that, "I am going on very slowly with my little drawings and shall not busy myself for most truly I have no spirit for work." Maria Bachman Audubon died on September 15, 1840, at the age of twenty-three.[7]

It was after Maria's death that the senior Audubon shouldered even more of the burden himself. No one has ever suggested that Audubon had a hand in producing the drawings for the octavo edition or that John did not do them all. But John was unable to keep up the pace, and Victor and Eliza, who suffered from the same malady that had felled her sister, soon left for Cuba in hopes that the tropical climate would speed Eliza's recovery. Thus Audubon had to do the reductions himself or see the most successful project of his career come to a halt. His travel journal is silent for late September and October of 1840, which he spent at home, probably making many of the small drawings that Bowen needed to continue the work. The journal is also silent from late December 1840 to July 11, 1842, during which time the family purchased land on the Hudson River and began construction of "Minnie's Land," the family home.[8] Audubon probably produced additional drawings during those months.

We know that the usually hardworking Audubon was working fourteen hours a day during this period, ostensibly on the paintings for the quadrupeds project but also probably on the octavo drawings from number 19 through number 68. His "Day Book" documents at least three instances in which he received payment of $7 per number, the same as he paid John, for making small drawings or "outlines": $98 on June 12, 1841 (Plate 34), for numbers 19 through 32 (plates 91 through 160); $105 on January 14, 1842, for numbers 33 through 47 (plates 161 through 235); and $147 on December 12, for numbers 48 to 68 (plates 236 through 340), a total of fifty numbers, 250 drawings, or half of the numbers in the octavo edition. When Parke Godwin, who had met him in the offices of the New York *Evening Post*, visited "Minnie's Land" in 1842, he saw "stuffed birds of every description of gay plumage" on "the mantle-piece; and exquisite drawings of field-mice, orioles, and wood-peckers, . . . scattered promiscuously in other parts of the room, across one end of which a long rude table was stretched to hold artist materials, scraps of drawing-paper, and immense folio volumes, filled with delicious paintings of birds taken in their native haunts."[9]

The change in style is apparent in the few extant original drawings now in the Hill Memorial Library, Louisiana State University. *Wilson's Flycatching-Warbler* (Bowen 75), the *Hemlock Warbler* (Plates 35 and 36), and the *Caerulean Wood-Warbler* (Bowen 86), the first three of the remaining drawings, apparently were drawn by John or perhaps redrawn by R. Trembley, since they are quite similar to the *Mourning Ground Warbler* (Plate 32), which is signed "R. T." in the lower left-hand corner. (Trembley drew most of the octavo edition birds on the stone, and Audubon must have been pleased with his work, for he rewarded him with a $25 bonus on New

Year's Day, 1841.) *Wilson's Flycatching-Warbler* and the *Hemlock Warbler* are even marked with a grid pattern, which might have been part of the effort to correct the distortions of the camera lucida or a part of the process of drawing them on the stone.[10]

By comparison, the *Hermit Wood-Warbler* (Plate 37), the first extant sketch in those numbers that are attributed to Audubon himself in the "Day Book," is a more sophisticated drawing with a surer sense of the birds themselves. Other drawings that seem to be by Audubon, rather than by John or by one of Bowen's artists, are the *Red-Eyed Vireo* (Bowen 243), the *Red-Bellied Nuthatch* (Bowen 248), the *Canadian Woodpecker* (Bowen 258), the *Downy Woodpecker* (Bowen 263), the *Yellow-Billed Cuckoo* (Bowen 275), only a fragment of which survives, the *Welcome Partridge* (Plate 40), the *Willow Ptarmigan* (Plates 38, 39), the *Sora Rail* (Plate 43), the *Virginian Rail* (Bowen 311), the *Whooping Crane* (Bowen 313), the *Rocky Mountain Plover* (Bowen 318), and the *Missouri Meadow Lark* (Bowen 489). The *Black-Bellied Plover* (Bowen 315) and *Harris' Finch* (Bowen 484) are detailed and handsome drawings but lack the lifelike look that is so characteristic of Audubon's work. The pencil and watercolor of *Townsend's Wood Warbler* (Bowen 92) in the collection of the Audubon Memorial Museum in Henderson, Kentucky, is a much more finished work. By comparison, the *Sandwich Tern* (Bowen 431), the *Ivory Gull* (Plate 42), the *Tufted Puffin* (Bowen 462), the *Curled-Crested Phaleris* (Bowen 467), the *Knobbed-Billed Phaleris* (Bowen 468), the *Slender-Billed Guillemot* (Bowen 475), *Bell's Vireo* (Bowen 485), the *Yellow-Bellied Flycatcher* (Bowen 490), the *Least Tern* (Bowen 491), and the *Western Shore-Lark* (Bowen 497) seem little more than carefully drawn outlines.

Audubon's drawings are not his usual finished work, and he even referred to them in one instance as "small outlines," but they are clearly better than the workmanlike efforts that characterize the other surviving drawings.[11] The facial features of the *Red-Eyed Vireo* (Bowen 243) seem particularly well defined, for example. The *Red-Bellied Nuthatch* (Bowen 248) and the *Downy Woodpecker* (Bowen 263) are among the most detailed and lifelike drawings in the surviving group, again suggesting a much more talented hand than either John or Bowen's copyists had shown. The calm little drawing of the *Sora Rail* (Plate 42) does little to call attention to itself—for the background is clearly unfinished—until you notice the female, at the left, looking directly at the viewer, an intimate gesture that only Audubon could have made convincing. The facial features of the *Virginian Rail* (Bowen 311) and the *Whooping Crane* (Bowen 313) are also distinctive.

The discovery of these drawings means that Audubon was even more intimately associated with the octavo edition than has been suggested, involving himself in the process, from conception to production of the drawings for the lithographers, and of the text for the printer; working with design and oversight of the book; and advertising and selling the finished product. This does not diminish anyone else's contribution to the book as much as it does help us understand that this team effort was still firmly under the guiding genius of Audubon himself.

Victor, for example, did not forget the project while en route to Cuba with his ailing wife. He paused in Mobile and New Orleans to sell subscriptions and to urge his father to increase the print runs to 1,100 lithographs and 2,000 of the letterpress. Shortly after his arrival in Havana, he wrote that he had sold Count Fernandina a subscription. In fact, subscriptions were coming in so rapidly that Audubon asked Victor to stop selling in Cuba and not to sell any bound volumes unless he received $100 payment in advance, because he could not fill all the orders he had in the United States and did not want to take on any far-flung obligations that might prove difficult or costly. To keep ahead of demand, Audubon ordered Bowen to increase the number of lithographs to 1,000 with number 11 and to 1,250 with number 24, leading Victor, whose dealings with Bowen were still fresh in his mind, to "hope Bowen will be able to get out the 1250 copies of the little work *well & regularly.*"[12]

Despite the fact that Eliza's health steadily worsened, Victor kept up with the business as best he could. On February 23 he provided his assessment of the numbers, probably 23 and 24, that he had recently received. "We . . . on the whole think them very fair, altho' I am inclined to believe the *graining* (in shading) is not as good as some of the previous numbers." He suggested that if Bowen could work from the original paintings, rather than the Havell prints, "the work w[ould] be so much improved. . . . The imitation of Havell's *aquatint* at present is disagreeable to the eye, and will be a great drawback to the value of the work if not put a stop to—do think seriously of these suggestions." When Eliza's health did not improve, she and Victor returned to New York, where she died on May 25, 1841, at the age of twenty-two.[13]

The frantic pace, in the meantime, had outstripped Bowen's production capability, and printing stalled at number 25. He was fully occupied in December 1840 and January 1841, producing a second printing of number 10, a third printing of numbers 7, 8, and 9, a fourth printing of numbers 5 and 6, and a fifth printing of numbers 1 through 4, all delivered in January 1841. (See Table 1.) In the midst of this rush, "the Stone on which were drawn *The Marsh* and *Winter Wrens*, broke after a few impressions and will put us off for several Days," Audubon reported. "All hands are now engaged in the bringing forth Back Numbers from 1 to 24. We cannot expect to supply any of the new Subscribers (about 200) with Setts before the 9th or 10th of next month. The fact is that my Success at Boston &c and yours at the South came upon us almost, if not quite unexpectedly." Later Audubon admitted, "It is a tremendous thing to bring up 24 Nos." No wonder he impatiently snapped at Little and Brown in April 1841: "I doubt much if you are *actually aware* that we have at this moment in this city and at Philadelphia upwards of *Seventy* persons employed upon the present work and that all these . . . are to be paid regularly each Saturday evening, and that *when we are out of temper* it is not without cause." By the time number 25 was finally delivered, it was forty-one days late.[14]

These production complexities have led to several oddities in some copies of the *Birds*. The title legends vary throughout according to the style and hand, going from block letters in the earliest numbers to Audubon's preference, italic script, in later numbers. Perhaps the block lettering, which was employed on the plates in the first three hundred copies of number 1, and the three incorrect titles in the first five hundred copies of number 5 can be used to date the earliest published copies of the plates for the octavo edition. Some prints are different from set to set, probably because a stone broke or was damaged and had to be redrawn, maybe by a different artist, or perhaps because Audubon changed his mind. An obvious example is the *Black-shouldered Elanus* (Bowen 16). The earlier version shows the two birds about 1/32 of an inch apart, in the same close relationship as in the Havell plate (352). A second state, which also appeared with the first edition text, shows the birds about 3/4 of an inch apart.[15]

The uneven number of prints and letterpress produced led to other anomalies as well, such as the binding of first printing of letterpress with later states of the lithographs. Audubon saved the extra letterpress sheets and bound them with the new lithographs to produce as many finished copies as he could, but making no attempt to match the first printing of the text with the first state of the lithographs. "In the Course of next week," Audubon informed Victor in January 1841, "we expect to receive a parcel of Back Numbers to No 10 Inclusive, and . . . we will *here* be able to make up about 15 Setts which will be sent off in a crack to the nearest hungry mouths from which we may expect Immediate remittances." As the publication drew to a close, however, the Audubons sometimes had trouble piecing together complete sets. "I hope you will not sell any more *odd* Nos of the little work," Victor informed Audubon in May 1844, because "they may be the very ones we want. However any below No 56 we can spare." The U.S. Department of Agriculture at one time owned a partial set of the *Birds* that contained the first edition letterpress and second state lithographs, which are distinguished by having tinted backgrounds and were not printed until 1856; so the Audubons apparently kept the extra letterpress on hand for years.[16]

The increased pace and size of the orders put Bowen in financial difficulty again. Some have suggested that Audubon might have paid him slowly, as he had Havell, compounding the lithographer's financial difficulties, but an examination of the "Day Book" suggests that the venture was profitable from the first and that Audubon paid promptly, if not according to the precise schedule that he and Bowen had agreed upon. The problem was the collapse of the McKenney and Hall project, which had reduced Bowen's capitalization. Audubon's orders were so large that Bowen could not carry the costs of paper and production until the lithographs were delivered, and now he apparently demanded payment from Audubon each week, claiming that he had already lost $800 on the job. He threatened to take his seventy lithographers and colorists off the task if Audubon did not comply. Audubon might have had less sympathy for Bowen if he had heard, as was rumored by the young George Burgess, that Bowen "had made money by his profession but lost all by speculating in stocks."[17]

Audubon also complained that Bowen could not keep up with the demand and that he had begun to simplify the backgrounds of some of the drawings, perhaps as a part of his effort to meet Audubon's demanding schedule. "About a week ago Mr. Bowen got on his high horse and went so far as to write to me that he would give up the work! Telling us big stories about his losses &c &c and complained bitterly that every number of Johny's drawings contained more work than the last," Audubon recalled,

> I wrote a letter to Bowen with the will and wishes of Mamma and Johny in which I told him that we were ready to accept his resignation and asking him to appoint a day for settlement and *actual* payment of the balance! Bowen was at breakfast with us the very next morning but one. We received him *as usual* extremely kindly; he showed us the coloured proofs he had brought along and I refused them at once and for ever. He stared not a little, but on his hearing me tell him that in case I should give up his engagement, that I would send Johny to England that very day for the purpose of bringing over 50 workmen as good as himself, he mellowed down as an apple does in an oven, and ere he left us the same day did promise us never to complain again and ask as a particular favour that I would burn his letter, which however I told him I would retain for the "Sake of Old Lang Syne," and I hope we will have no further trouble with him for *a good while*.

> The fact was simply this, that he had taken upon himself to *cut up our little drawings at such a rate*, that I was determined to check him, and I have done [so] effectively. Chevalier is properly delighted and so are we all. He has furnished Chevalier with back *Nos* up to 9 inclusive but it will be something like 2 months before he comes up to No 24.[18]

Although Bowen probably did not take seriously Audubon's threat to send John to England for colorists to finish the job, he did know that other lithographers were available. Perhaps he had heard that Audubon had again approached Henry Havell, who was continuing the family business in New York, and had received a bid that was $2 less per hundred lithographs than Bowen was charging.[19]

Audubon finally made good on his threat to employ another lithographer in February, probably to give Bowen more time to catch up on the back numbers, but clearly a message to Bowen that the job could be done elsewhere. He employed George Endicott of New York to lithograph fifteen hundred copies of the fifteen images in numbers 28, 29, and 30, the most he would print of any of the illustrations, and contracted with J. W. Childs, also of New York, to color them. Nor did the work with Endicott go flawlessly, as a colored proof before lettering of the *Common Mocking Bird* (Plates 46, 47), now in the collection of the Hill Memorial Library, shows. Small differences between this image and the published version suggest that Audubon rejected the proof and had the image redrawn.[20]

It is difficult to decide whether Audubon's charges against Bowen were justified. Only a few of the drawings for the first twenty-four numbers, over which the dispute arose, are known to exist, and a comparison with the Havell engravings is not as meaningful as one might assume because there are considerable differences between the extant octavo drawings and the Havell plates. Audubon made a number of changes when he produced the small drawings and whatever differences there are between the Havell plates and the Bowen plates might have been his own doing. The drawings of *Wilson's Flycatching-Warbler* (Bowen 75, Havell 124), the *Hemlock Warbler* (Bowen 83, Havell 134), and the *Caerulean Wood-Warbler* (Bowen 86, Havell 48) are similar to the Havell prints, except that the sprigs of turtlehead, mountain maple, and dahoon, respectively, have been much simplified in the reduced format. Bowen has simplified them even further in his prints, suggesting that there might have been some substance to Audubon's complaints. But if the drawings did not exist, and one were comparing the Havell plates with the Bowen plates, it would be easy to conclude that Bowen had made substantial changes. The *Hermit Wood-Warbler* (Bowen 93), the other extant drawing for these early images, is quite different from the Havell plate (Havell 395), which contains three species and six different birds. Audubon gave each species a plate to itself in the octavo edition; the other four birds from the Havell plate are illustrated as *Audubon's Wood-Warbler* (Bowen 77) and *Black-throated Grey Wood Warbler* (Bowen 94). There are many other examples of Bowen plates that feature simplified or deleted backgrounds when compared to the Havell plates—for example, the *Red-tailed Buzzard* (Bowen 7, Havell 51), the *Golden Eagle* (Bowen 12, Havell 181), the *White-Headed Sea Eagle, or Bald Eagle* (Bowen 14, Havell 31), and the *Gos Hawk* (Bowen 23, Havell 141), to list only a few; but the octavo drawings for these plates are not known to exist.

On the other hand, Bowen, like Havell, assisted Audubon with a number of the drawings. The landscape background in the print of the *Whooping Crane* (Bowen 313), for example, is much more elaborate than in the drawing. At the bottom of the drawing of the *Welcome Partridge* (Plates 40, 41), Victor wrote: "Put a Landscape to these plates of Partridges." And Bowen obliged with a handsome little habitat complete with a fallen tree trunk.

Bowen, of course, had compelling reasons for remaining in Audubon's employ. The miniature birds were among the finest natural history illustrations produced in this country—surely the best reproductions of Audubon's paintings other than the original double elephant folio up to that moment. Bowen was proud of his contribution to such a prestigious project and exhibited the prints as examples of his work. Shortly after the octavo edition was under way, Audubon began work on *The Viviparous Quadrupeds of North America*, an even more ambitious project, and Bowen wanted that job, too.[21]

The royal octavo edition of the *Birds* became the most popular natural history book in America during a period of severe economic distress. The panics of 1837 and 1839 initiated a depression, from which the nation did not emerge until the mid-1840s. A glance at Audubon's account books, and the different cash accounts that he maintained in New York and Philadelphia, reminds one of the unstable economic conditions that marked these decades. Audubon kept money in several different banks, not only for the convenience of collecting and paying bills and supporting family and employees in both cities, but also to prevent being wiped out if one of them collapsed. "Think of the [$2,000] We have as dead in Kentucky Bank Stock," he advised Victor at one point. To complicate the matter, Audubon had trouble with Chevalier, who apparently was unable to keep up with the project. One of the first hints of difficulty arose when Victor warned Audubon, in March 1840, "*Do not* mention these matters to C[hevalie]r when you write to him." Toward the end of the year, as Victor was en route to Cuba, he commented that John would have difficulty balancing Chevalier's books, suggesting that Audubon had sent John to Philadelphia to get to the source of the problem. The following January, Audubon advised Victor that, in addition to Bowen's problems with McKenney and Hall, "the Bank of the U.S. is very low *at present* and in all prob-[ability] will have to wind up."[22]

Audubon's subscriber base was also threatened by the unsteady economy. "We have had great havoc among the banks of Phil[adelphia,] Baltimore and Virginia, all of whom almost Simultaneously suspended species payments," he advised Victor in February. "I am also afraid that all those rascally banks' difficulties will affect our procuring new Subscribers in the Spring when I contemplated making a great Sweep at New York, Albany, Troy, Rochelle, Quebec, &c &c &c. . . ." When Victor returned from Cuba, he promptly went to Philadelphia to see if he could resolve any remaining questions with Chevalier. "Brother and Mr. Chevalier are now hard at the books," John wrote in June. Victor added in a postscript: "It will be a matter of *time & difficulty* to close with *J. B. C. even if necessary*." When Audubon visited Chevalier in February 1842, he found him "not a little frightened at the present aspect of affairs [and] concluded to stop the *extra* printing of letterpress beyond that wanted for the Nos called for at present or supposed to be wanted." They also lowered the price of binding each number in an effort to encourage new subscribers to purchase the back issues, thereby selling the "immense pile" of letterpress that they had in stock. Shortly thereafter, Chevalier severed his ties with the Audubons and sailed for France to join his family. Victor advised in July 1843, "I have some trouble with our money matters, but hope to get along tolerably." As Audubon tried to bring the project to a close, he needed to collect from subscribers who had not yet paid and to find out how many copies his agents had, particularly the large ones like Little and Brown, who handled quite a few copies. Victor wrote in July 1844 that "Little & Brown have 474 Nos of the small work on hand which brings us in their debt considerably!"[23]

As volume 5 drew to a close, John had returned to his task of drawing the little birds, while the developing *Viviparous Quadrupeds of North America* demanded more of Audubon's time. "I am now as anxious about the publication of the *Quadrupeds* as I ever was in the procuring of our Birds—indeed my present interest in Zoology is altogether bent toward the completion of this department of natural science," Audubon had written to the young Spencer F. Baird, who would later serve as assistant secretary of the Smithsonian Institution, in July 1841. His 1843 trip up the Missouri River to Fort Union was for the purpose of gathering further species for the *Quadrupeds*, but he found "no less than 14 New Species of Birds, perhaps a few more," he wrote Bachman upon his return. "The variety of Quadrupeds is small in the Country we visited, and I fear that I have not more than 3 or 4 New ones." Audubon busied himself with paintings for the *Quadrupeds*, but he also apparently redrew some of his new bird paintings for inclusion in the "little work." Among the extant drawings that seem certain to be his work is the *Missouri Meadow Lark* (Plate 50).[24]

The miniature edition of *Birds of America* was complete by 1844. It corrects several errors that Audubon had made in the double elephant folio, such as those that he addressed through the production of the thirteen composite plates. It contains seven species of birds that were described in the *Ornithological Biography* and listed in the *Synopsis* but not illustrated in the double elephant folio and seventeen birds that were neither illustrated nor described in the earlier works. Volume seven, for example, includes a lithograph of the *Texan Turtle Dove* (Bowen 496), the only bird included in Audubon's work that is directly traceable to a Texas specimen. Audubon reported that the skin had been supplied to him by J. G. Bell, who received it from an unnamed correspondent in Texas.[25]

Some of the octavo plates show whimsical additions, such as the small sailboat in the lower left-hand corner of the picture of the bald eagle (Bowen 13, Havell 11), an addition that Audubon had made to at least one oil painting of the eagle before he began the octavo edition.[26] A number of Havell plates were redrawn to show only one bird or species to a plate. Plate 434, illustrating the little tyrant flycatcher, the blue mountain warbler, the short-legged pewee, the small-headed flycatcher, Bartram's vireo, and the Rocky Mountain flycatcher, for example, was divided into seven Bowen prints: the *Rocky Mountain Flycatcher* (Bowen 60), the *Short-legged Pewit Flycatcher* (Bowen 61), the *Least Pewee Flycatcher* (Bowen 66), the *Small-headed Flycatcher* (Bowen 67), the *Blue Mountain Warbler* (Bowen 98), *Bartram's Vireo or Greenlet* (Bowen 242), and the *Least Flycatcher* (Bowen 491). Havell plate 417 (Plate 44) was similarly divided into six octavo plates (Bowen 258, 259 [Plate 45], 260, 261, 265, and 269); plate 416 was divided into five octavo plates (Bowen 262, 266, 270, 272, and 274); and plates 362, 394, 400, 402, 424, and 432 were divided into four each (228, 230, 232, and 235; 154, 182, 193, and 194; 153, 157, 178, and 183; 467, 468, 470, and 471; 187, 197, 198, and 207; and 29, 30, 31, and 38). Some of the other plates were divided into two or three Bowen prints. In other instances, such as the *Goshawk* (Havell 141), the plate was re-

32. R. Trembly after John James Audubon, *Mourning Ground Warbler*, 1840. Pencil and red pencil, 6 13/16 by 3 1/2 in. (image). Courtesy Hill Memorial Library, Louisiana State University. Bowen employed Trembly to copy most of Audubon's images on the stone. Perhaps because John Woodhouse Audubon had submitted an inferior drawing, Trembly drew the mourning ground warbler (reproduced as Bowen plate 101).

33. W. after John James Audubon, *Black-Throated Wax Wing*, 1842. Pencil and brown pencil, 7 by 4 1/2 in. (image). Inscribed: "Philada Jan 10- 1842." Courtesy Hill Memorial Library, Louisiana State University. "W." probably was one of Bowen's artists. Unlike most of the known octavo edition drawings, this one is not the same size as the print. There are some slight differences in size and configuration that might be credited to the use of a camera lucida.

34. John James Audubon, "Birds of America Day Book," page 120, 1841. Ledger. By permission of the Houghton Library, Harvard University. This page from the "Birds of America Day Book" shows that Audubon first received payment for doing the "small drawings" on June 12, 1841. (Despite the date at the head of the ledger, entries for June 12 begin on the previous page.) He began with number 19. He received other payments for the small drawings on January 14 and December 12, 1842.

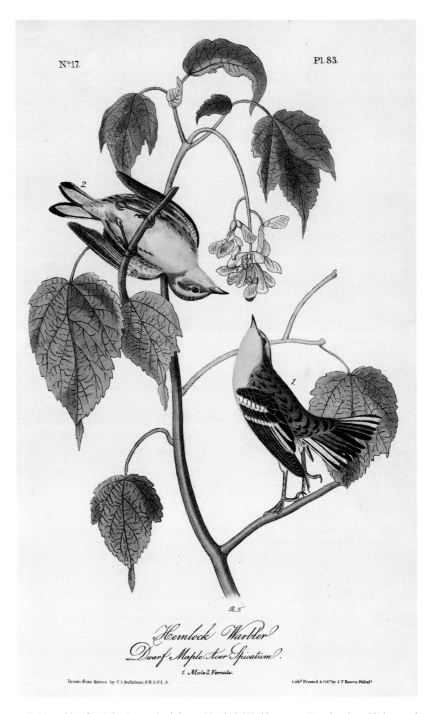

35. R. Trembly after John James Audubon, *Hemlock Warbler*, 1840. Hand-colored lithograph, 10 1/4 by 6 1/2 in. (sheet). From *The Birds of America*, plate 83. Courtesy W. Thomas Taylor, Austin, Texas.

36. John Woodhouse Audubon (attrib.) after John James Audubon, *Hemlock Warbler*, 1840. Pencil and red pencil, 9 by 5 11/16 in. (image). Courtesy Hill Memorial Library, Louisiana State University. Most of the octavo plates are much simplified compared to the double elephant folio. The *Hemlock Warbler* is no exception, with fewer leaves on the dwarf maple in the octavo plate.

37. John James Audubon, *Hermit Wood-Warbler*, 1840. Pencil, 8 3/8 by 5 3/16 in. (image). Courtesy Hill Memorial Library, Louisiana State University. This is the first drawing in the collection that is by Audubon. The difference in style alone suggests that—the leaves and branch of the strawberry tree are more detailed and sophisticated than the other "outlines"—and references in the "Day Book" support that conclusion. The drawing has been rubbed on the back, suggesting that it was transferred to the stone for lithographing, but it is slightly larger than the lithograph used for comparison, raising the possibility that there might have been another drawing and another state of the print. This drawing was taken from plate 395 of the double elephant folio.

38. John James Audubon, *Willow Ptarmigan*, 1840. Pencil and brown pencil, 4 7/16 by 7 9/16 in. (image). Courtesy Hill Memorial Library, Louisiana State University. The drawing of the male (left) is slightly larger than the lithographed image (plate 49). The hints of shrubs in the background do not appear in the print. The drawing has been rubbed on the back, and there are wax spots on the face.

39. J. C. after John James Audubon, *Willow Ptarmigan*, 1842. Hand-colored lithograph by Bowen, 6 1/2 by 10 1/4 in. (sheet). From *The Birds of America*, plate 299. Courtesy W. Thomas Taylor, Austin, Texas.

40. John James Audubon, *Welcome Partridge*, 1842. Pencil and brown pencil, 3 1/8 by 6 in. (image). Courtesy Hill Memorial Library, Louisiana State University.

41. J. C. after John James Audubon, *Welcome Partridge*, 1842. Hand-colored lithograph by Bowen, 6 1/2 by 10 1/4 in. (sheet). From *The Birds of America*, plate 292. Courtesy Hill Memorial Library, Louisiana State University. Victor Audubon wrote at the bottom of the drawing, "Put a Landscape to these plates of Partridges—" J. C., Bowen's artist, obliged by adding a landscape with a dead log, which is only hinted at in the drawing.

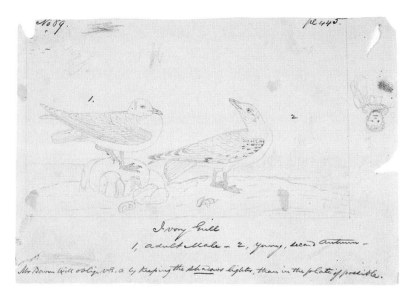

42. John Woodhouse Audubon after John James Audubon, *Ivory Gull*, 1843. Pencil and brown pencil, 3 by 8 1/4 in. (image). Courtesy Hill Memorial Library, Louisiana State University. The small landscape suggested by the drawing at the right did not make it into the finished print (Bowen 445). Victor Audubon provided instructions at the bottom: "Mr Bowen will oblige V.G.A. by keeping the *shadows* lighter, than in the plate if possible." He must have been referring to the Havell plate 287, which seems to have heavy shadows.

43. John James Audubon, *Sora Rail*, 1842. Pencil and brown pencil, 4 3/8 by 8 1/4 in. (image). Courtesy Hill Memorial Library, Louisiana State University. The published version is virtually identical to the drawing except that the background has been enhanced. The drawing of the female (left) is indicative of Audubon's ability with a gesture or a look.

44. Robert Havell, Jr., after John James Audubon, *Maria's Woodpecker. Three-toed Woodpecker. Phillips Woodpecker. Canadian Woodpecker. Harris's Woodpecker. Audubon's Woodpecker*, 1838. Hand-colored aquatint and engraving, 30 1/2 by 22 1/2 in. (plate). From *The Birds of America*, plate 417. Courtesy Stark Museum of Art, Orange, Texas. Audubon began to crowd several species together toward the end of his project. He had already extended the publication date beyond what he had promised and feared that he was trying the patience of his subscribers.

45. J. C. after John James Audubon, *Phillips Woodpecker*, 1842. Hand-colored lithograph by Bowen, 10 1/4 by 6 1/2 in. (sheet). From *The Birds of America*, plate 259. Courtesy Edward Kenney. In the octavo edition, Audubon illustrated each species on a page by itself. He made six octavo plates similar to this one out of Havell plate 417.

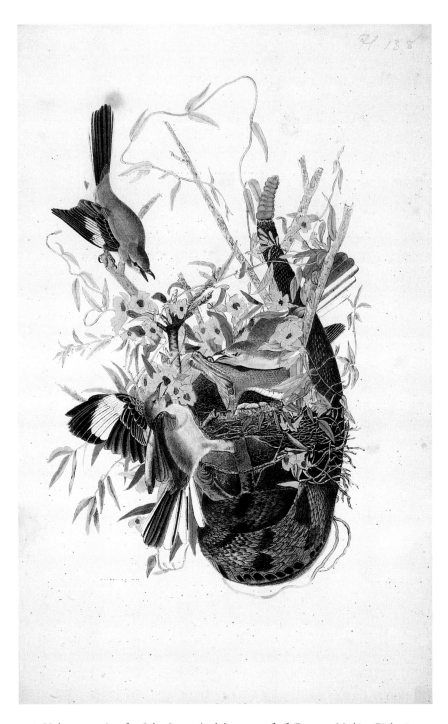

46. Unknown artist after John James Audubon, proof of *Common Mocking Bird*, 1841. Hand-colored lithograph by G. W. Endicott, 10 3/4 by 6 7/8 in. For *The Birds of America*, plate 138. Courtesy Hill Memorial Library, Louisiana State University. Audubon employed George Endicott of New York City to produce numbers 28-30 of the octavo edition. Several changes were made, as may be seen by comparing this proof before lettering of the mockingbird to the finished plate (right).

47. Unknown artist after John James Audubon, *Common Mocking Bird*, 1841. Hand-colored lithograph by G. W. Endicott, 10 1/4 by 6 1/2 in. (sheet). From *The Birds of America*, plate 138. Courtesy W. Thomas Taylor, Austin, Texas. Minor changes were made throughout; the plate seems to have been completely redrawn.

48. R. Trembly after John James Audubon, *Arkansaw Flycatcher*, 1840. Hand-colored lithograph by Bowen, 8 3/4 by 5 1/16 in. (plate). From *The Birds of America*, plate 54. Courtesy W. Thomas Taylor, Austin, Texas.

49. R. Trembly after John James Audubon, *Arkansaw Flycatcher*, 1856. Hand-colored, tinted lithograph by Bowen, 8 15/16 by 5 1/8 in. (plate). From *The Birds of America*, plate 51 (second state). Courtesy W. Thomas Taylor, Austin, Texas. These two Arkansaw flycatchers, which both should have had "a patch of bright vermillion on top of the head," show why Audubon was often upset with Bowen. Only the second state (plate 48) is accurately colored.

50. John James Audubon, *Missouri Meadow Lark*, 1844. Pencil, 5 1/2 by 5 1/8 in. (image). Courtesy Hill Memorial Library, Louisiana State University. This is one of the western birds that Audubon discovered on his trip up the Missouri River in 1843. The drawing has been rubbed on the back and has wax spots on the face. It was reproduced as Bowen plate 489.

drawn to correct some of Havell's compositional errors and bad judgments. John presented a simplified drawing for the octavo edition that looked more like Audubon's pasted-up original than Havell's engraving (Bowen 23; *Goshawk*, O.P. 17).[27]

Virtually all of the crowded or complex Havell prints have been simplified in the octavo version so as to preserve the aesthetic appearance that Audubon valued, but because both Bowen and the Audubons themselves hurried to finish the task, some of the octavo plates may have been simplified beyond what Audubon wanted. Early in the project Bowen complained that the drawings were too complex, and Audubon threatened to take the job elsewhere if Bowen did not stop editing them himself. The *Sparrow Falcon* (Havell 142, Bowen 22), in which several leaves and branches have been cut, and the *American Redstart* (Havell 40, Bowen 68), in which many of the leaves have been omitted, may be examples of images that Bowen simplified himself. The landscapes in the backgrounds of the octavo drawings of the *Canadian Woodpecker* (Bowen 258), the *Common Tern* (Bowen 433), the *Ivory Gull* (Bowen 445), and *Harris' Finch* (Bowen 484), and the hints of shrubs in the background of the *Willow Ptarmigan* (Bowen 299), do not appear in the prints. Also, the leaves and flowers in the drawings of the *Downy Woodpecker* (Bowen 263) and the fragment of the *Yellow-Billed Cuckoo* (Bowen 275) have been reduced in the print. Surely Audubon himself cut the squirrel out of the illustration of the *Barred Owl* (Havell 46, Bowen 36). Some of the more complicated landscapes, such as Charleston in the background of the *Long-Billed Curlew* (Havell 231) and the swamps behind the *Roseate Spoonbill* (Havell 321), were simplified, probably because of the impossibility of getting as much into the reduction (Bowen 355 and 362). As a result, these landscapes do not seem as integral a part of the picture in the octavo edition as they do in the folio.

The text of the octavo *Birds* is a reprint of most of Audubon's *Ornithological Biography*, with the sections that he called "delineations of American scenery and manners" omitted. Because the total number of subscribers on Audubon's published list is over 1,200, several authors have estimated the number of complete copies printed and bound of the first octavo edition at about 1,200. But, as with the double elephant folio, not all those who subscribed completed their subscriptions. Audubon had reduced the print run on the plates to 1,150 beginning with number 51 and to 1,050 beginning with number 57, so there could not have been 1,200 subscribers at the finish, and he could not have produced 1,200 bound copies of the first edition without a reprint of the plates that is not included in his "Day Book." (He apparently had sufficient copies of the text.) The total number of subscribers might have been nearer 1,000 or 1,050. The hundred numbers were issued over a period of almost five years, beginning in 1839 and concluding in 1844. (See Table 2.)

It has often been said that Audubon's real popularity did not begin until the octavo edition of the *Birds* was published. While that is probably not true, he did greatly expand his audience with the new edition, for it was published at the height of Americans' frenzy for "national" works of art and literature. Audubon believed that his subscribers consisted of men and women of "liberality" and "taste." It goes without saying that they must also have been wealthy. Boston Brahmins and Louisiana planters bought his books, as did learned societies, libraries, the old rich, and the new mercantile class. On his subscription list were senators and congressmen, cabinet members, military officers, lawyers, doctors, editors, merchants, and artists. Naturalists John Cassin and Spencer F. Baird subscribed, as did New Orleans merchant Germain Musson, the grandfather of French artist Edgar Degas, and a Mrs. Bruce, who ran a boardinghouse on Tchoupitoulas Street in the Crescent City. Many of these subscribers also patronized worthy causes such as public charities and churches, and some saw Audubon's books—indeed, Audubon himself—in this same category, because his work was science and natural history, or because it was a great national, i.e., American, work. The *Saturday Courier* called it "an honor to the American nation," the *Albion*, a "national and instructing work." "In times like the present," the editor concluded, "this is highly honourable to parties who thus munificently encourage science."[28]

Audubon had designed his publications to appeal both to the scientifically minded and to those who admired the fine arts. These apparently divergent fields were much more complementary in pre–Civil War America than today. A number of the subscribers, such as Robert Gilmor, Jr., of Baltimore, patronized American artists, while others served on the boards of arts and natural history organizations, and three of them were artists themselves (including Robert Havell, Jr., who purchased two subscriptions). Such people joined the societies, contributed money, and on occasion, researched and published books themselves. Most of Audubon's subscribers accumulated libraries.[29]

Although Bachman urged him to ransack the countryside, most of his customers were in the five largest cities in the country: New York City (142, or 11.5 percent), Boston (207, or 16.7 percent), Philadelphia (74, or 5.9 percent), Baltimore (168, or 13.5 percent), and New Orleans (61, or 4.9 percent). (See Table 3.) Charleston, while not as large as New Orleans, was a hotbed of naturalists and provided Audubon with 69 subscriptions (5.6 percent), followed by Washington, D.C., and New Bedford (49 each, or 3.9 percent) and Richmond (41, or 3.3 percent). Massachusetts accounted for almost 30 percent (362) of his subscriptions, Maryland and New York had more than 14 percent (181 and 176, respectively) each, Pennsylvania almost 9 percent (108), and South Carolina a little more than 7 percent (90). Audubon received nineteen subscriptions (.015 percent) from Canada, twelve (.009 percent) from England, and two (.002 percent) each from China and Cuba.

This trend toward the coastal states and cities follows other economic trends in pre–Civil War America, when the easiest method of transportation, both for Audubon on his sales trips and for later fulfillment of his orders, was by water. It is also predictable when one considers the occupations of those who purchased the octavo *Birds*. A comparison between New York City and New Orleans suggests that the new merchant and middle classes strongly supported Audubon in both the North and the South. (See Table 4.) New York City provided a total of 142 subscriptions, New Orleans, 61. A total of fifty-four names (38 percent) in New York and sixteen (26 percent) in New Orleans remain unidentified, but of those who can be identified, the largest category of purchasers in both cities was merchants, thirty-two (23 percent) in New York and fifteen (25 percent) in New Orleans, for a total of forty-seven (23 percent) in both cities. They were primarily commission merchants in New Orleans, where the cotton trade dominated; there was a wider spread in New York, including dry goods and other retail merchants. The next largest category in New York was lawyers (seventeen, or 12 percent); in New Orleans it was planters (eight, or 13 percent). Five lawyers (8 percent) subscribed in New Orleans. The third largest categories in New York were bankers and doctors (six each, or 4 percent), while doctors and government employees followed in New Orleans (three each, or 5 percent). Three artists (2 percent) and a taxidermist (less than 1 percent) subscribed in New York. In the capital of Washington, D.C., on the other hand, out of a total of forty-nine subscribers, at least twenty-two (45 percent) were associated with the military, four (8 percent) were employed by the government, two (4 percent) were in the Cabinet, and one (2 percent) was a U.S. senator. Three doctors (6 percent) and one editor (2 percent) are also identifiable in Washington, but no merchants. (See Table 5.)

While it is difficult to be sure how much money Audubon earned, he might have made as much from the octavo edition as from the double elephant folio but in less than half the time. He fulfilled his promise of one hundred fascicles of five colored prints each for $100. He also offered the set of seven volumes bound in half or full morocco leather for $115 and $120. He generally paid his agents a 10 to 20 percent commission, so he could expect to receive from $.80 to $.90 from the sale of each number, more if he himself sold it. If his final expenses totaled an estimated $.60 to $.65 per number, he could expect to receive approximately $.25 to $.35 profit per number. If he sold all of the 1,050 sets printed, he would have had a potential profit of $25 to $35 per set (100 numbers x $.25 or $.35 per number). Since he sold 85 percent of the sets himself, his profit could have been as much as $35,980 ($35 x 896 sets = $31,360; to this, add $30 [average income on agent's sales] x 154 sets = $4,620), the equivalent of $563,555 in 1991. He would also have received some income from subscribers who did not complete their subscription; he printed 1,500 copies of the lithographs for numbers 28 through 30, for example, and might well have sold more than 1,050 copies of them. Furthermore, profit from the octavo edition would have been spread over a four- to five-year period rather than an eleven-year period, as it was with the double elephant folio.[30]

Reviews of the work were uniformly good. From England the knowledgeable William Yarrell, author of *A History of British Fishes* and *A History of British Birds*, wrote of the small birds that "I like them much—as I could not afford to have the large work I make myself content with the small one.... I am quite of your opinion that there would be some sale for it here—if it was advertised and made known." A German visitor praised Bowen for the "beautiful lithographic work and . . . brilliant lifelike coloring." The *Saturday Courier* quoted a Professor Wilson, "one of the ripest scholars of Europe," as saying that Audubon's literary style "is the most natural, flowing, easy, graceful, and, at the same time, touching, terse and truly eloquent, of any with which I am acquainted. I wish I could write with so much naivete, expression, freedom, and perfect picturing of the object to be presented to the mind's eye. But that style is, in part at least, but the inspirings of God." The *Albion* concluded that his text "breathes poetry in every line."[31]

Several months later, the *Saturday Courier* continued its coverage of Audubon's work, predicting that his "Birds of America, and . . . Quadrupeds of America will be great National Works, which will live to glorify his talents and perseverance so long as a love of Natural history shall endure." Poets praised his accomplishments in the press, and his bird stories were excerpted in the *Albion, Godey's, Brother Jonathan, Southern Cabinet*, and *Parley's Magazine*, among others. Some, like the *Albion, Arthur's Magazine*, and A. A. Gould in *A Naturalist's Library* (Boston, 1849), published engraved adaptations of several of his images. Even if it is true that few Americans would have seen a folio or an octavo plate, most of them had read about the "celebrated naturalist" and many would even have read one of his stories.[32] The fullness of Audubon's genius was at last brought home for the American public to enjoy. And in so doing he had ensured his family's financial security for the remainder of his life and much of his sons' and his wife's.

CHAPTER FIVE

Subsequent Editions of "The Birds of America"

Audubon turned his full attention to the *Viviparous Quadrupeds of North America* even before the octavo edition of *Birds* had been finished. He was putting in fourteen-hour days on the project as early as 1842, and the Reverend Bachman, who was to be co-author of the text, anxiously urged him on. Audubon had threatened Bowen with taking the work to someone else and had talked with Henry Havell about it, but few lithographers in the country were turning out the quality of work that Bowen routinely produced. Imperial-size lithographs (22 by 28 inches) of the *Quadrupeds* finally began to appear from Bowen's press in early 1843, even before Audubon set out on his Missouri River expedition to gather additional specimens for the work. Victor pronounced them "good," but Bachman was more generous, calling the prints the "most beautiful and perfect specimens of the art" that he had ever seen (Plate 66).[1]

With Victor managing the business affairs, as usual, John producing paintings and drawings, and Bachman writing the text, the senior Audubon took off on his only far western trip in the spring of 1843, an expedition that he had planned for literally decades but had been unable to arrange. It almost came too late for both Audubon and the West. "Mr. Audubon is quite an aged man," wrote one of the reporters along the route, "but his active and hardy life has given a vigor and strength to his constitution which renders him far more active than the generality of men of his years."

While in Saint Louis, Audubon visited with old friends and turned down Sir William Drummond Stewart's offer of a wagon and five mules if he would join his party en route to the last fur trapper rendezvous in the Wind River Mountains. Finally, accompanied by longtime friend Edward Harris, who was preparing a report on the geology of the upper Missouri for the Philadelphia Academy of Sciences, and Isaac Sprague, a young artist whom he had met in Massachusetts, Audubon boarded the *Omega* on April 25 and steamed up the Missouri. The trip seemed more of an adventure than a research expedition, with Audubon being received along the way as a celebrity. They debarked at Fort Union in June, and Audubon and Harris occupied the same chamber that had housed Prince Maximilian and Karl Bodmer ten years before. Audubon enjoyed the buffalo hunts (which he vividly described in his journal) but ultimately found very little in the way of new quadrupeds.[2]

The West, too, had changed greatly in the intervening decades. Had he been able to see it during the 1820s, when he first attempted to go, Audubon might have seen more of the landscapes or the noble savage that he hoped for. When he saw the Assiniboins, whom he said artist George Catlin had represented as "dressed in magnificent attire, with all sorts of extravagant accoutrements," he found them, instead, "very dirty." Catlin had also misrepresented the landscape, he said. "We have seen much remarkably handsome scenery, but nothing at all comparing with Catlin's descriptions; his book must, after all, be altogether a humbug." Audubon arrived back at Minnie's Land in November, in time to include the fourteen new species of birds that he had discovered on the expedition in the last volume of the octavo edition.[3]

When Victor triumphantly reported that number 100 of the *Birds* had been delivered on June 21, 1844, the family turned its full attention to selling the remaining copies of the small edition and to full-time production of the *Quadrupeds*. While John made a trip to Texas looking for additional quadrupeds, Victor nudged Bachman a bit in March 1846, writing, "We will be very glad to get the *Letterpress* from you, and I hope it will soon come to hand." At the same time he confessed, "We have I fear not sent you all the information we ought to have procured, but we have done as well as we could." That would have been a considerable understatement to Bachman, who complained, begged, and cajoled continually in an effort to get the Audubons to send him books, articles, and skins that he could use in writing the text. And to say that he, who knew little of the western quadrupeds and had been depending upon Audubon for the field work, was disappointed in Audubon's western trip is putting it mildly. When Bachman finally visited Minnie's Land in the fall of 1845, he sadly realized the problem: Audubon was becoming senile, and his sons had to work without his help—indeed, work around him in some instances.[4]

The imperial edition of the animals was finished in three volumes the following year. It contained 150 hand-colored lithographs. Without Audubon's active participation, the accompanying text, written mainly by Reverend Bachman and Victor Audubon, using Audubon's journals, was finally published in three separate octavo volumes under the same title, appearing in 1846, 1851, and 1854. Counting those subscriptions that

Audubon sold before his death in 1851, Victor and John probably wound up with a few more than three hundred subscribers.[5]

Even before the imperial edition plates were finished, Victor and John had been thinking about the miniature edition of *Quadrupeds*. The concluding text volume of the imperial folio, published in 1854, was also the beginning of the octavo edition, for it contained a supplement which included six hand-colored, octavo-size lithographs of animals that had been left out of the imperial folio. When the octavo edition itself appeared that same year, another new picture had been added: the *Mexican Marmot Squirrel* (Octavo 109), a combination of the two imperial folio plates of the *Mexican Ground Squirrel* (Imperial 109 and 124). The *Mountain Brook Mink*, one of the six new plates, filled the vacancy (Octavo 124) created by the merging of two ground squirrel plates, and the remaining five new plates were added at the end, making up Octavos 151–155. The Audubons sold perhaps as many as two thousand copies of the octavo *Quadrupeds*.[6]

Seventy-seven of the plates in the imperial folio edition reproduce Audubon's drawings; seventy-three of them are copied after John's work. The seven new plates are entirely John's work, the original drawings being made after his father's death. One only has to compare them with work that John did while his father was alive to realize how influential the senior Audubon was in the production of both the *Birds* and the *Quadrupeds*, regardless of whose name is on the plate as artist. The *Jackall Fox* (Octavo 151), for example, has none of the natural grace and elegance of the *American Black or Silver Fox* (Octavo 116) or the *Arctic Fox* (Octavo 121). Nor is the plate of *Col. Abert's Squirrel* and the *California Grey Squirrel* (Octavo 153) as convincing as *Collie's Squirrel* (Octavo 104) or the *Dusky Squirrel* (Octavo 117). These last five plates, produced by John, are easily the poorest compositions in the book and a vivid testimony to Audubon's controlling genius.

An 1853 letter that Victor wrote Bachman from Vicksburg suggests that the Audubons had continued to reprint the octavo *Birds* as needed, even while they were producing the *Quadrupeds*: "I have ... sold one Copy of the Large Birds and 7 copies of the small birds which will have to be printed and coloured." This practice probably led to a number of hybrid copies of the set. The fact that varying numbers of lithographs and letterpress were printed from time to time means that at any given moment the Audubons would have had differing quantities of pictures and text on hand to be made up into sets whenever they received an order. When they ran short of either, they ordered a new printing, as Victor suggested was necessary in this instance, without indicating, either on the letterpress or the lithographs, that they were later printings.[7]

In June 1851, John Cassin (Plate 51), a Philadelphia ornithologist, approached the brothers about a "continuation of the American Ornithology." Apparently what he had in mind was a set of ten volumes, the first seven a revision of Audubon and the last three to be his own work. Victor discussed the matter with John in March 1852, warning that they were still at work on the octavo animals and might not have time to do a new edition when "we *ought to be* going after subscriptions to the quadrupeds." Besides, Cassin wanted to include birds that had been discovered since the first edition had been published. "*If we are* to do the *new birds*," Victor warned, "we cannot make the first drawing for more than a year—and you will have to do nearly all of them in Philadelphia—& the *plants* all over the Country. We could not begin to publish before Jan[uary] 1854!" In response to subsequent letters from Cassin, Victor replied, "It would afford my Brother and myself great pleasure to join you in the continuation of my Father's work on the Birds, and we will gladly make an arrangement with you for that purpose—provided we can agree with you on terms."[8]

Cassin apparently proposed that he revise Audubon's nomenclature "in accordance with received rules," get $50 per month after the publication of number 20 ("I do not wish to wait until the completion of the work before I can get a penny"), and retain one-third copyright. No doubt the brothers thought this too much. "There are some points in your proposition to us that are not to our mind," Victor answered, "and . . . it seems to me we can only arrange the matter satisfactorily all round, if at all, by a *personal* understanding." He insisted that they retain Audubon's nomenclature, that he and John do "portions" of the text, and that the pictures include "plants, trees &c. from the localities where the birds are found," just as Audubon had always done.[9]

At the heart of Cassin's concern for the nomenclature was the lack of respect that professional ornithologists had for Audubon's and his sons' scientific work. Audubon made many mistakes in naming birds, of course, but there were more serious errors. George Ord of the Academy of Natural Sciences and others had quarreled with Audubon over various details in his writings from the moment of their publication. Some became causes that Audubon aficionados still discuss: Can rattlesnakes climb trees? Do vultures find their food more by sight or smell? In an 1832 review of the *Ornithological Biography*, the French naturalist Baron Cuvier, whose comment on Audubon's contribution to science is widely quoted, provided a more candid, professional appraisal: "Monsieur Audubon is not, however, a naturalist. He is a skillful painter and an intelligent observer." Ord summed up the matter after an encounter with Audubon in

1845: "If the fidelity of his narrative had corresponded with his perseverance, his fame would repose on a basis which time would not diminish, but what will be the decision of posterity on the merits of one who has wantonly violated the dignity of truths by rendering her subordinate to fiction!" Cassin felt the same way and could not afford to be associated with Audubon's work unless he were given the authority to make many corrections. Yet, he realized, as did Victor, that Audubon's work had flair and appeal that most natural histories lacked: "The cool manner in which *Baird and Co* have set down what's what—in regard to Birds, Beasts & reptiles is to me quite rich!" Victor wrote Bachman.[10]

Cassin launched a trial balloon early in 1852. "I intend to get up two volumes supplementary to Audubon's octavo edition and have the prospectus ready to print," he wrote colleague Spencer Baird, who was now assistant secretary of the Smithsonian Institution. Hoping to secure enough subscribers to finance the project, he issued a prospectus with Audubon's name in larger type than his own and described his work in later titles as "Descriptions and Figures of all North American Birds not given by Former American Authors."

Rather than hand-colored lithography, he could have chosen chromolithography, then offered by several American printers, as the method of creating his images. Chromolithography is a related process in which most of the colors are printed from additional stones, with only the finishing touches being applied by hand. Lithographers in New York, Philadelphia, Boston, and Baltimore had turned to it with the encouragement of the government, to illustrate several of its publications during the 1850s, such as Schoolcraft's *Historical and Statistical Information Respecting the History, Condition and Prospects of the Indian Tribes of the United States* and Emory's *United States and Mexican Boundary Survey*. Based on the chromolithographs produced for Emory, the cost was slightly less than Bowen and Company was charging for hand-colored prints: $.06 versus $.075.

Cassin would have known of Peter S. Duval of Philadelphia and James T. Ackerman of New York, who had just completed a series of chromos for Schoolcraft's history, and Lewis N. Rosenthal of Philadelphia, who was then at work on Alfred Jacob Miller's western illustrations for Charles W. Webber's sporting books. But chromolithographers could not approach Bowen's colorists in quality, so Cassin chose the more expensive method. He selected Henry Louis Stephens as the artist—an illustrator better known for his cartoons and birdlike caricatures of prominent individuals than for birds themselves—and approached Bowen about producing the lithographs. Swamped with

Audubon's *Quadrupeds* and other business, Bowen declined. Cassin then contracted with Rosenthal. Cassin, no doubt, would have sympathized with Audubon in his bouts with Bowen because of the ensuing aesthetic conflict he had with Rosenthal. "Stephens has nearly got out the first three plates. They are beautiful—birds and plants like Audubon," he wrote, but "I have to most carefully watch all hands and expect to have to fight either the lithographic printer who is the most obstreperous, or a young lady colorist who is not much better." His first number appeared to little fanfare.[11]

Sale by subscription was a difficult process, at best, in pre–Civil War America. Some publishers merely chose that method as one of several available to them to sell books that they had published, but others used it to raise money necessary to print expensive books. In such cases, it was helpful if not mandatory to have the personality and/or the compelling product of an Audubon to be successful. These Cassin clearly did not have, and his venture was unsuccessful. He listed only sixty-nine subscribers in his original prospectus. When no second number was forthcoming, those few who had subscribed wondered what had happened. "What progress is Cassin making with the continuation of Amer[ican] Ornithol[ogy]?" Jared P. Kirtland, a charter subscriber, asked Baird. John Woodhouse Audubon's verdict probably represented the most common point of view: "Cassin's first No. of the birds is very bad." Cassin realized his error, as well, admitting to Baird that "I have made a serious mistake in getting out my first Number with such indifferent plates."[12]

The experience, no doubt, led Cassin to continue his pursuit of the Audubon brothers. "I saw Cassin and Harris" in Philadelphia, Victor related to Bachman. "I do not know yet whether we shall attempt to publish the new Birds in connection with Cassin or not; he is to be here this week to see me about it." Two weeks later, Victor concluded that "our negotiation with Cassin is I suppose broken off, but—I shall know by tomorrow—he is too anxious to do something however to let the matter pass and may go on by himself."[13]

Cassin, in the meantime, had found a new artist, George Gorgas White; a new printer, Bowen; and a publisher, J. B. Lippincott of Philadelphia, to help with subscriptions and to distribute the book, and he issued his work in parts beginning in March 1853, almost a year after his first effort. "The Prospectus for the present Work was issued so long since as January, 1852," he explained to his subscribers, "and a small Edition of a first part was published in April of the same year. Since that time it has become practicable to engage the most accomplished and experienced Lithographic Artists . . . in the country. . . . [T]he Publisher and the Author . . . have deemed it expedient, therefore, to issue a second Edition of the first part, with new and greatly improved Plates."[14]

While White's drawings are slightly better than Stephens's, most of the improvement in the plates probably can be credited to William Hitchcock, one of Bowen's lithographic artists who had drawn many of Audubon's birds on stone. Still, few of Cassin's birds will be confused with Audubon's, for they are stiffly perched or standing on the ground and appear to be sitting for their portraits in comparison to Audubon's animated and intense creations. Cassin eventually became disenchanted with White and turned entirely to Hitchcock, letting him draw the birds directly on the stone without producing a finished watercolor as a model. This also saved a bit of money, something that Cassin had to think about, for he admitted in June 1854, "With my book, it is now a struggle to save myself from losing money."[15]

Nor was Cassin's text as readable as Audubon's. It was more scientific, to be sure; his apparently was the first American work to employ trinomial nomenclature, and he included dozens of birds not represented in Audubon, western birds that Audubon had not had a chance to see, such as the roadrunner, or ground cuckoo, as Cassin called it. Lippincott had distributed ten numbers with five plates each by the end of 1855. Then the entire set was bound with a new title page dated 1856 and offered to the public in one volume.[16]

Cassin knew before the ten numbers were completed that he had not obtained enough subscribers for the work to be a success. Knowing also that he still possessed enough material not in Audubon for two additional volumes, he contacted the Audubons again in mid-1855 and proposed that they publish a "fusion edition" of ten volumes, with his being the last three. "Let it be represented that the work is completely brought up to the state of ornithology at present as a complete work on the Birds of the U.S.," he wrote Victor in June. He still insisted that he be permitted to revise the nomenclature, but revised his other terms to include 10 percent of the sales and a quarter of the copyright. The Audubons still did not agree. Cassin had intended to publish three volumes, but when the proposed cooperative endeavor fell apart, he knew that he would not be able to complete them. His book was eventually reprinted twice, in 1862 and 1865, but the print runs were so small that probably no more than a few hundred copies were ever published, and it is an extremely rare book today.[17]

Probably one reason that Cassin's volume enjoyed no better sales was that the Audubons published a second edition of their small *Birds* in 1856. The letterpress text and images, with few exceptions, are the same as the first edition. In some cases, in fact,

Subsequent Editions

they are the first edition. The Audubons probably had quite a bit of letterpress still on hand, since they had printed from one thousand to two thousand copies of each text until the changing economic climate led Audubon to reduce the print run in 1842. They also had kept the stereotype plates and stones from the first printing, a result of Audubon's early warning to Victor to keep the plates in his possession, as Audubon had kept the copperplates from the Havell edition of the double elephant folio. This was more unusual with lithography, because the stones could easily be ground smooth and reused, just as the copper in the double elephant folio plates could have been melted down and reused. Victor had written Audubon in 1843 of an attempt that he had made to sell the stones, but the sale apparently was not completed. The Audubons kept them and reprinted the plates as needed during the years following completion of the first edition; then, in 1856, they reprinted the entire work.[18]

While the text is the same, the illustrations for this edition are easily distinguished from those of the first, because Bowen added a blue-green tint to the background to every plate except those with landscape backgrounds. The tint was printed separately from the image with another lithographic stone, rather than painted by hand (Plates 52, 53). Bowen's staff prepared this stone to print a solid color around the original image of the bird and limb, habitat, or foreground. It was probably printed first, then the image. The image was then hand-colored, just as with the illustrations for the first edition. A proof of the tint stone for *Swainson's Swamp Warbler* (Plate 53), now in the collections of the Hill Memorial Library, clearly documents the procedure.

In addition, there are a number of differences between the first and second state plates other than just the background tint. Most of the title legends had to be redone because they had worn virtually off the stone. The artists' initials were also badly worn and in many cases not replaced. Most of the changes to the images are just as minor, such as in the *Californian Turkey Vulture* (Bowen 1), where the limbs and trunk on which the bird perches appear to be slightly cropped in the second state. This, too, could be explained by a worn lithographic stone, by a variation in inking or in hand coloring, or perhaps some other eccentricity of the lithographic process. Other differences are more significant, such as in the illustration of the *Belted Kingfisher* (Bowen 255), in which the hill in the background of the first state has been replaced with the tint coloring in the second. Surely some of the lithographs had to be made over for the same reasons that Audubon had encountered during production of the first edition: breaking stones and editorial changes. A plate-by-plate comparison of the first and

Subsequent Editions

second state plates suggests that a number of them might even have been redrawn, because the images in the second state seem slightly larger—1/32 to 1/16 of an inch from the top of the image to the bottom—than those in the first state. These include *Harris's Buzzard* (Bowen 5), *Harlan's Buzzard* (Bowen 8), the *Broad-Winged Buzzard* (Bowen 10), and the *White-Headed Sea Eagle, or Bald Eagle* (Bowen 14), among others. In the final analysis, however, most of the first and second state plates appear to be identical insofar as the birds and the essential parts of the image are concerned.[19]

The octavo edition of *Quadrupeds of North America* was reprinted in 1854 and 1856, while both it and *Birds* occupied preeminent positions in their fields. The writer and philosopher Henry David Thoreau noted in his journal on June 25, 1856, for example, that he went by a local library to identify a bird that he had seen in a friend's yard in Audubon's "hundred dollar edition." In 1858 the U.S. Secretary of State was authorized to purchase one hundred copies of each as gifts to foreign governments, unofficial acknowledgment of the national role that Audubon's works had assumed.[20]

In a twist of fate, Cassin soon became involved in the subsequent editions of Audubon's books. As a result of his own book, Cassin realized the enormous possibilities for influencing the scientific as well as aesthetic content of publications by controlling the illustrations. He felt that he had gained sufficient experience on his own book to become more involved in the process and had written to Baird in an effort to secure a job as a sort of supervisor for the illustrations accompanying the mammoth *Pacific Railroad Reports*, which the government published between 1855 and 1860. "I have some idea of offering to contract for them myself—that is in connection with Bowen,—I find that both Bowen and Hitchcock have to be directed and supervised," he told Baird, "and I think I might as well if I can get into a position of some authority with them.... I think that I will offer proposals for myself & Bowen—making Bowen pay me a small commission if possible." Although Cassin did not get the *Pacific Railroad Reports* job, the relationship with Bowen proved to be mutually beneficial. They won several contracts through Cassin's natural history and government contacts, which, in turn, enhanced his reputation among his scientific peers.[21]

The relationship with Bowen's firm soon became more formal. Following Bowen's death in 1856, his widow needed Cassin's business acumen and managerial ability as well as his contacts to maintain the company. He worked closely with her, finally becoming half-owner and president of the company in the spring of 1858. His influence in the natural history world had never been greater, and Bowen and Company became preeminent in the field. When John returned to Bowen and Company in 1859

Subsequent Editions

to issue a third edition of *Birds of America,* Cassin produced it, an irony that he probably savored.²²

Although the text of the third edition is virtually the same as that of previous editions, with the exception of the publisher and copyright information on the title page and verso, a number of the plates have been changed dramatically, and one is tempted to wonder if Cassin had anything to do with the changes. The third edition was co-published according to an agreement that John reached with Roe Lockwood and Son of New York, probably because Victor had suffered a spinal injury in a fall in 1857 and no longer could manage the family's affairs. The tints behind most of the birds remain, as in the second state, but in some cases the background was changed. The *Californian Turkey Vulture* (Plates 54, 55), shown perched on a limb in the first two states, for example, has been placed on a cliff overlooking a valley in the 1859 state. The *Black Vulture or Carrion Crow* (Bowen 3) has been redrawn somewhat larger, and a landscape with small vultures sitting in trees at the left has been added. A horizon line has been added to the plate of the *White-Headed Sea Eagle, or Bald Eagle* (Bowen 14); a range of mountains has been added at the left, the eagle's left claw now rests on a branch, a fish head has been added at the lower right center, and the attribution to Bowen has been replaced with a credit line to Hitchcock.²³

Some of these changes might have been made to the original stone, but in other cases, the plates have been completely redrawn. The *Burrowing Day-Owl* (Plates 56, 57) is one of the most fascinating. The first and second states are a rather uninteresting vertical illustration of the male and female, one on the ground and the other shown as a vignette perched on a suspended branch. The male has obviously spotted some prey, for his beak is open, as if calling, and he is looking rather intently toward the lower left. In the third state, the image has been made horizontal with the two owls on the ground and a desert landscape behind them. This is probably a more accurate habitat for the bird, for Audubon reported that "its predilection for the ground forms a very distinctive peculiarity" of its habits. He said that he had never had the opportunity of seeing it procure food, but that an examination of its waste suggested that small quadrupeds made up a large portion of its diet. The change in this illustration reflects John's hand. He had gone to California in 1849 to get in on the gold rush and had made a number of drawings, which were intended as an *aide-mémoire* for paintings and prints to be made later. The giant saguaro cactus and the dead tree shown in this landscape come directly from one of his drawings labeled "Sept–12, 1849—14 miles N.W. of Altar," Sonora, Mexico (Plate 58). "We have seen this cactus for about 5 days travel, say 125 miles," he wrote at the bottom of the drawing. "Woodpecker's hole in top is Centurus uropygialis Baird, 1858." It is a handsome landscape, but hardly squares with the information that Audubon had obtained from Thomas Say about the birds being natives of the "plains near the Columbia River and the whole extent of the Rocky Mountains."[24]

The changes to the *Barn Owl* (Bowen 34) are modest by comparison and might not have required the production of an entirely new stone. In the first two states, the male and female owls are shown frolicking on their tree limbs set against a beautiful night sky. Their wings are spread and faces animated in anticipation of the upcoming feast on a squirrel, which is held in the talons of the bird at the upper left. Since they are nocturnal, the background of the plate is both tinted and hand-colored in some copies, giving it a rich, dark color that suggests that the owls are out in the dark of the night. In the third state, the background has been lightened to suggest twilight, and the profile of a cityscape is silhouetted in the background across the lower part of the plate. The owls seem to be in the same position on their limbs.

Equally as different as the two states of the *Burrowing Day-Owl* are the first and third states of the *Short-eared Owl* (Bowen 38). The third state is a horizontal plate with the owl on the ground by a stream, rather than a vertical illustration with the owl on a vignetted tree limb, as in the first and second states. A full landscape has been added to the third state, including a turtle on a log in the stream. The result is a much more appealing picture than the original print. Other significant changes in the 1859 state include the *Belted Kingfisher* (Bowen 255), which has also been changed from a vertical to a horizontal plate, and the birds rearranged in the new space. The small bird that was perched on a stump at the left has been placed on a rock in the middle of a stream. The landscape has been completely redrawn and brought closer to the foreground, so that the bird on the stump, which appeared to be quite in perspective with the distant background, now appears to be huge and out of context with the new, closer landscape. The mountains have been taken out of the landscape, and the stump, banks, and forest completely redrawn.

Some of the changes probably were not intentional. The *Great American White Egret* (Bowen 370), the *Reddish Egret* (Bowen 371), the *Snowy Heron* (Bowen 374), the *American Flamingo* (Bowen 375), and the *Sandwich Tern* (Bowen 431) appear to have been redrawn by an inferior artist with no attribution. Several parts of the images, such as the grasses, have been simplified, and there is a general loss of detail in the birds' feathers. Perhaps the lithographers were trying to rework greatly worn stones. On the

51. Unknown photographer, *John Cassin*, c. 1860. Cabinet photograph. Prints and Photographs Division, Library of Congress. One of the outstanding ornithologists of his day, Cassin wanted to work with the Audubon brothers to produce a revised edition of *The Birds of America*. When they could not reach agreement, he proceeded with his book and the Audubons with a second edition of theirs.

52. R. Trembly after John James Audubon, *Swainson's Swamp Warbler*, 1856. Hand-colored, tinted lithograph by Bowen, 10 3/4 by 6 7/8 in. From *The Birds of America*, plate 104 (second state). Courtesy W. Thomas Taylor, Austin, Texas. Bowen added a tint block to almost all of the birds in the 1856 edition. The only exceptions were those that had landscape backgrounds.

53. Tint block for *Swainson's Swamp Warbler*, 1856. Lithographed by Bowen, 10 3/4 by 6 7/8 in. Prepared for *The Birds of America*, plate 104 (second state). Courtesy Hill Memorial Library, Louisiana State University. The lithograph was printed in two colors, black for the image and blue-green for the tint, and then hand colored. This is the tint block for Swainson's warbler, showing that space has been left for printing the bird, the azalea, and the butterfly.

54. R. Trembly after John James Audubon, *Californian Turkey Vulture*, 1839. Hand-colored lithograph by Bowen, 10 1/4 by 6 1/2 in. (sheet). From *The Birds of America*, plate 1. Courtesy Stark Museum of Art, Orange, Texas.

55. R. Trembly after John James Audubon, *Californian Turkey Vulture*, 1859. Hand-colored lithograph by Bowen, 10 1/4 by 6 1/2 in. From *The Birds of America*, plate 1 (third state). Courtesy DeGolyer Library, Southern Methodist University, Dallas, Texas. In the 1859 edition, John and Victor made a number of changes to enhance certain plates. They added, for example, a handsome landscape to the plate of the turkey vulture.

56. R. Trembly after John James Audubon, *Burrowing Day-Owl*, 1840. Hand-colored lithograph by Bowen, 10 3/4 by 6 7/8 in. (sheet). From *The Birds of America*, plate 31. Courtesy W. Thomas Taylor, Austin, Texas.

57. Unknown artist after John Woodhouse Audubon, *Burrowing Day-Owl*, 1859 or later. Hand-colored, tinted lithograph by Bowen, 5 13/16 by 8 13/16 in. From *The Birds of America*, plate 31 (third state). Courtesy DeGolyer Library, Southern Methodist University, Dallas, Texas. The burrowing day-owl was placed in a desert landscape, based on John's 1849 drawing made near Altar, Sonora, Mexico.

58. John Woodhouse Audubon, *14 Miles N.W. of Altar*, 1849. Pencil, 10 1/4 by 13 in. Courtesy Southwest Museum, Pasadena, California. John wrote in his notes that the hole in the top of the cactus was made by a woodpecker.

59. Unknown artist after John James Audubon, *Fish Hawk*, 1890. Chromolithograph, 6 7/8 by 4 5/8 in. (image). From Warren, *Birds of Pennsylvania*, plate 80. Courtesy William S. Reese, New Haven, Connecticut. Besides diminishing the anthropomorphism of Audubon's birds, Warren did not show them high in the air, as Audubon had. Here the fish hawk perches on a rock with the weak fish still in its claws.

other hand, both plates of the *Wild Turkey* (cock and hen, Bowen 287 and 288) were also redrawn on a new stone by Max Rosenthal, a well-known Philadelphia lithographer, probably because the old stone broke or wore out. There are minor differences between the states, but, in this instance, at least, the new artist took credit for his work.[25]

Given Cassin's long-held desire to make some changes in Audubon's work, the question as to how much influence he might have had in these revisions may be resolved in the fact that the cactus and dead tree in the lithograph of the *Burrowing Day-Owl* originated in John Woodhouse Audubon's drawing. This suggests that the family was still in control of the book.

About a year after Victor's accident and spinal injury, and as soon as the third octavo edition was finished, John undertook a full-size reprint of *Birds of America* by chromolithography. This was possible because the family still had almost all of the original copperplates that Havell had engraved, and because John joined with New York publisher Roe Lockwood and Son, the same distributors that he had successfully worked with on the third octavo edition, to handle the sales and distribution. As he prepared to return from England in 1839, Audubon had carefully packed and shipped the copperplates to New York. A few of them had to be restored "to their wonted former existence" after a disastrous fire destroyed an entire section of the city in 1845, including the warehouse where they were stored, but the family had preserved them in "the Cave," which they constructed on a slope behind the Audubon home in 1852 and 1853.[26]

John made arrangements with Julius Bien and Company, 180 Broadway in New York, an established lithographer and map engraver who had immigrated to the United States from Germany in 1849, to do the printing. Bien had attended the Academy of Fine Arts in Kassel as well as the Städel Art Institute in Frankfurt am Main and was one of the many German "forty-eighters" who came to America to start over. He employed two hundred people and fifteen steam lithographic presses, along with related machinery, in a successful business and had received many awards for the maps that he engraved for the *Pacific Railroad Reports*, the census reports, and various U.S. Geological Survey atlases and military field maps. Like Havell in 1826, Bien was a specialist at the top of his form in 1858. He agreed to provide John 250 copies of each number for $1,250 (additional copies at the cost of $5 per number). John agreed to pay him one-half the amount due for each number in cash, the other half to be financed by a three-month note, which was to pay interest at the rate of 7 percent per annum.[27]

John planned to reproduce all of the original double elephant folio prints in forty-four numbers, each containing ten images on seven sheets (two large images, two medium images, and three sheets containing two smaller images each). A forty-fifth number would contain the text. The prospectus for the English trade, issued in 1859, predicted a "softness, finish, and correctness of coloring, [that] will be superior to the first, and every Plate will be colored from the original Drawings, still in possession of the family. . . . The first Number [*Wild Turkey* cock] is considered superior in many respects to the same Plates in the first Edition, and it is confidently hoped that subsequent Numbers will exhibit still greater superiority as the Artists gain experience." The cost was to be £2.8 per number, or approximately $11, which would have totaled about $495, including the text, a considerable savings over the cost of the Havell edition, yet providing a healthy profit for the family. No copy of the prospectus for the American trade has been located.[28]

Only because he would be doubling up on the smaller images and because of significant advances in printing technology during the intervening twenty years could Bien undertake to duplicate Havell's feat at less than half the cost. He planned to capture the subtlety of Havell's engraving and aquatint by transferring the images from the copperplates to lithographic stones. The process by which this was to be accomplished had been in use virtually since the development of lithography in 1798 and had been considerably refined by the time Bien undertook the Audubon project. It required that Bien ink the original copperplate, cover it with a thin, dampened sheet of transfer paper, and run it through a copperplate printing press. A sharp, accurate image of Audubon's bird, rendered in black ink, was the result. The paper was then laid face down on a prepared stone and run through a lithographic press, transferring the image to the stone. A number of transfer inks and papers were available to Bien. After additional work on the stone to fix the image and prepare it for printing, a piece of blank lithographic printing paper was laid on top of it, and it was run through the press again. The result was a duplicate of the engraving, except that it had been printed by lithography, which was less expensive than printing engraving and aquatints and could yield more copies. All the tonal qualities of the aquatint remained.[29]

The next step was coloring the print. This was perhaps Bien's greatest accomplishment. First, he had to decide how many colors he would use to duplicate the range of colors in Havell's print. Then he had to select the inks, decide the order in which they would be printed, and figure out how to keep each subsequent impression in register with those already printed. Seldom using more than six colors, Bien captured the tonal ranges of Havell's colorists by varying shades, juxtaposing certain colors, and laying one color over the other. In some instances, he finished off the print with touches of hand-coloring, indicating that chromolithography was not yet fully mechanized. It was a monumental achievement, involving a more sophisticated understanding of the process and technical control than any similar contemporary undertaking.[30]

Despite his best effort, however, it is clear that only in his finest work did Bien approach the quality that Havell uniformly achieved. Even the *Wild Turkey* cock (Havell 1, Bien 287), which John probably had Bien print as an experiment and held up as representative of the proposed set, lacks the rich color and the fineness of detail that are hallmarks of the hand-colored aquatint. At times the results are much less desirable. The chromolithographs vary significantly from print to print, so generalizations are dangerous, but a comparison of Bien's *Greenshank* (Bien 346) with Havell's (Havell 269), for example, shows that the subtle and delicate watercolors have been replaced by jarringly bright blue-greens and orange-yellows that are today more associated with the later pre-Raphaelites and turn-of-the-century modernists than with Audubon or nature.

The problem, of course, was the proper mix of colors. This most demanding of tasks was the job of the chromistes, "master lithographers . . . whose chromatic perception is so acute that they can tell you at a glance what the great Turner himself did not know—how many colours go to the making of one of Turner's pictures." The chromistes analyzed each picture to be reproduced, drew each of the separate stones that would make up the image, determined the order in which the colors would be printed, and of course, checked the finished prints for correctness. The Bien chromos were the finest reproductions that could be obtained at half the price of the original edition, but they could not match the blend of colors and sensitivity of a gifted artist like Audubon—at least not at this early date.[31]

The seven-volume octavo edition text, bound in five volumes, was reprinted without illustrations in 1861, to accompany the Bien edition. Three copies of this edition are known to exist with colored frontispieces but no other plates; there might well be others.[32]

The chromos might well have found more of a market had the Civil War not intervened. Collectors were probably reluctant to make a substantial long-term investment in such troubled times. Philadelphia bookseller William Brotherhead noticed that by 1860 many of his southern customers were already having difficulty paying their bills. By the following year, he recalled: "Business was paralyzed, war, war, was the cry everywhere, all business except that of war was thrown aside. . . . Newspaper literature took the place of books. No one had any time to read except war news, and amid all this excitement for one year books were forgotten.... In 1862 war literature was fairly commenced and all other literature was subordinated to it." After the Union navy blockaded southern ports, the Audubons would have been unable to get in touch with their southern subscribers to deliver prints, even if they had been interested enough to make the investment. John later blamed the "effects of our distracted country" for his worsening financial condition.[33]

The project also may have failed because of some kind of financial mismanagement among the partners. John was never the businessman of the family. The senior Audubon had been disappointed in the results of John's trip to Texas in 1846 in search of new animals for the quadrupeds project and came to doubt his ability to carry through on any project, commenting, "He has never kept a plan in view for more than a few days." After Victor's injury, according to John's daughter, "All the care of both families devolved on my father. Never a 'business man,' saddened by his brother's condition, and utterly unable to manage, at the same time, a fairly large estate, the publication of two illustrated works, every plate of which he felt he must personally examine, the securing of subscribers and the financial condition of everything—what wonder he rapidly aged, what wonder that the burden was overwhelming. After my uncle's death [in 1860] matters became still more difficult to handle."[34]

The expenditures for the 105 chromos that were produced (and for printing the text that accompanied them) completed John's financial ruin and seriously affected Mrs. Audubon's fortunes as well. In an effort to avert disaster, the family sold its double elephant folio, the quadrupeds, and several other books to John Taylor Johnson, one of the founders of the Metropolitan Museum of Art in New York, in 1861, for $1,000; the double elephant folio is today in the Stark Museum of Art. "Worn out in body and spirit, overburdened with anxieties, saddened by the condition of his country," John's daughter continued, "it is no matter of surprise that my father could not throw off a heavy cold which attacked him early in 1862." He was only forty-nine when he died. Mrs. Audubon was forced to sell the original watercolors for *Birds of America* in 1863. The New-York Historical Society raised $4,000 by public subscription to purchase them. The breakup of the double elephant folio materials was completed in 1870 when she sold the surviving copperplates. They were sold and resold, some perhaps being contributed to a World War II scrap drive. As many as seventy-eight of the plates are known to survive today. The bulk of the Bien prints were offered for sale in 1883 by the Boston firm of Estes and Lauriat.[35]

Subsequent Editions

Meanwhile, the royal octavo edition of *Birds of America* continued to enjoy great popularity as George R. Lockwood and Son reprinted it in 1865 and 1870. These editions, bound in eight volumes rather than seven, were still printed, where possible, from the same stones and stereotype plates that had been made in the 1840s and 1850s. There had been a number of changes, as various stones were broken or damaged, but the book remained essentially as Audubon had issued it in the 1840s. Any subsequent editions would have to be made from completely new plates because as Richard B. Lockwood explained sometime after 1870,

The lithographed stones from which the plates were printed were destroyed by falling thru floors in a Bldg in Phil[adelphia] where they were stored—so that no more plates can be printed. Sets of Audubon in the Folio original are very rare, and bring high prices $3750 and up. This present Ed[ition] can be picked up now and then, while the prices increase . . . but any sets so secured are by necessity "secondhand" and it would be advisable for any one securing a set at a fair price to hold on to it as they are getting harder to secure each year.[36]

There are a number of references to an 1889 edition. (See Appendix.)

A partial copy of the octavo edition appeared in 1890 when the state of Pennsylvania included ninety-nine chromolithographic copies of Audubon prints in B. H. Warren's *Report on the Birds of Pennsylvania*. "The greater part of the illustrations in this report have been copied (some alterations as to positions, etc., have been made in nearly all) from the small edition of 'Audubon's Birds of America,'" an anonymous writer explained. "By copying from said work which is regarded by competent critics as containing many of the finest portrayals of birds that have ever been published, the cost (ten to twenty-five dollars per plate) of original drawings was saved."[37] While chromolithographers and their adherents inevitably claimed that their prints were superior to the hand-colored originals, that was not often the case. Nor is it in this instance. The chromos have a darker, and sometimes richer, color, but they possess none of the brilliance and subtlety so common in Bowen's work. Some might go so far as to suggest that the colors are not as accurate as Audubon's, either.

Some of Warren's changes were intentional. The large birds, for example, are pictured standing up, rather than in the contorted poses that Audubon had copied from his double elephant folio. There was no reason for him to have retained the awkward pose of the *Great Blue Heron* in the double elephant folio (Havell 211, Bowen 369), except that he liked it and wanted the identity with the double elephant folio, for the bird is not presented life-size in the octavo edition. In the *Birds of Pennsylvania*, the heron is shown in plate 69 standing up, rather than stooping to pick up a morsel. Several of the chromos have been simplified in other ways, sometimes omitting a bird but usually reducing the foliage and flowers that are so characteristic of Audubon's best work. Finally, Warren, given the advantage of decades of scientific work, must have influenced the lithographic artist(s) to reduce the anthropomorphic qualities so abundant in Audubon's work and to produce more objective bird portraits. Neither the *Great Horned Owl* (Bowen 39, Warren 19) nor the *Fish Hawk* (Plate 59), for example, possesses the ferocity that Audubon endowed them with; even the weakfish, which seems stunned at its fate in Audubon's plate, has become inanimate in Warren's representation.

One thing stands out about the Audubons in retrospect: Every enterprise was undertaken as a family, even after the senior Audubon died in 1851. The family joined Audubon in England in 1830 and helped him finish the double elephant folio in 1838. Upon their return to America in 1839, they threw themselves into production of the small *Birds* and the imperial folio edition of the animals even before the *Birds* was finished. Then, as soon as the imperial edition of the *Quadrupeds* was completed in 1848, Victor and John began the octavo edition in 1849 without their father's assistance, completing it in 1854. It proved to be so popular that they began issuing a second printing of volumes one and two even before volume three of the first printing had been completed. And as soon as the small edition of *Quadrupeds* was done, they began work on a second edition of the small birds, reprinting it in 1859 and 1860, by which time John was involved in the chromolithograph edition with Bien.

None of this would have been possible without the guiding genius of John James Audubon, who produced paintings of stunning quality, guided the printed copies through the successive presses of Havell and Bowen, and successfully marketed his "Great Work" to an important segment of American society. If the attempt to do great work is one of the hallmarks of genius, Audubon cannot be faulted, for the very comprehensiveness of his task—the attempt to catalog, illustrate, and describe all the birds and quadrupeds of North America at a time when many of them were still unknown to science—distinguished him from his peers, many of whom achieved noteworthy successes in their own right. The real test of genius, however, is in accomplishment, and in that his feat remains unparalleled.

CHAPTER SIX

Audubon in American Art

Despite his enormous accomplishments, his popularity, and the fact that more has been written on him than on most American artists, Audubon is seldom considered in the context of his painterly contemporaries and has never been properly recognized as one of the great American Romantic artists. His followers generally emphasize his tremendous accomplishments and the scientific aspects of his bird and animal portraits. They claim aesthetic merit for his paintings, but ultimately see them as natural history rather than fine art. His detractors pay even less attention, choosing to emphasize his lack of formal training in both science and art. As the artist George Catlin observed, Audubon's "works would seem to hold a rank between living nature and art." In fact, his paintings combine elements of contemporary portrait, genre, and landscape painting that any Romantic would have recognized, and his writings elaborate upon and deepen that context.[1]

By the time of his second return visit to America in 1831, with many of the double elephant folio engravings published and the first volume of the *Ornithological Biography* in print, Audubon was a well-known and popular figure. His huge aquatint engravings were well reviewed in *Blackwood's*, *Literary Gazette*, and *Silliman's*—"the Papers here have *blown me up* sky high," he wrote Lucy—the American Philosophical Society and the Academy of Natural Sciences in Philadelphia subscribed to his book, and he was invited to membership in the learned societies in this country and Great Britain. He exhibited his paintings and prints in galleries on both sides of the Atlantic and visited with artists and collectors, journalists and naturalists, congressmen and senators, and presidents and heads of state. He was the sort of celebrity whose comings and goings were noted by the press. But when success came his way, it was generally at the hands of the public rather than from naturalists or critics, although by the end of his brilliant career, he had friends in both worlds.[2]

Perhaps the main reason Audubon received so little attention from the critics and the art establishment is that they saw him as a natural history illustrator. Almost everyone admitted the spontaneity, creativity, and liveliness of his paintings, but in the end, he was a self-trained artist-naturalist whose work most critics saw as being entirely outside the academic tradition. They did not think of him in the same context with Doughty, Cole, Durand, Allston, Mount, or other contemporaries. Painter and historian William Dunlap thought Audubon a significant enough figure to include in his *History of the Rise and Progress of the Arts of Design in the United States* in 1834 but reported in detail his rejection by the natural history community in Philadelphia. Also, art historian Henry Tuckerman, whose biographical sketch of Audubon appeared shortly after the artist's death, recognized that his "career as an ornithologist began and was prosecuted with an artistic rather than a scientific enthusiasm" and that "the popular basis of Audubon's renown, as well as the individuality of his taste as a naturalist, rests upon artistic merit." However, when it came time to summarize his research into the artist's life a decade and a half later, Tuckerman reduced his account of Audubon to less than a paragraph, with no analysis of his work or suggestion of its lasting qualities.[3]

Others have suggested that Audubon might be little celebrated as an artist because he never became proficient at oil painting, the preferred medium of fine artists and collectors. He attempted oil paintings on occasion and produced a number of them to exhibit and sell in Britain, but he never invested the time and effort necessary to master the technique, ultimately only dabbling with it. Instead, he engaged the youthful Joseph Kidd to copy his watercolors in the more permanent format. He himself was much more at home with watercolors and used them throughout his life.

It also might be said that Audubon's residence in England during production of the double elephant folio kept him from identifying with or becoming a part of the generation of artists that developed into America's first native school of painting, the Hudson River School. Audubon had departed for England in 1826, shortly after landscapist Thomas Cole, the acknowledged inspiration of the school, made his first sketching trip up the Hudson; and Audubon had been in England for the better part of ten years by the time Cole painted his masterpiece, *View from Mount Holyoke, Northampton, Massachusetts, after a Thunderstorm (The Oxbow)* (1836, oil on canvas, 51 ½ by 76 inches, Metropolitan Museum of Art, New York). The naturalist did not move back to this country until 1839, by which time the most creative part of his work was finished, and the second generation of the Hudson River School was well established.[4]

Audubon fares little better with art historians today. Barbara Novak, for example, mentions him only in comparing his interest in birds to Martin Johnson Heade's fascination with hummingbirds. James Thomas Flexner compares him passingly to Catlin, and Milton W. Brown and Oliver W. Larkin do no better. John Wilmerding is one of the few art historians to mention him in his own right, observing that "he is usually set apart from the mainstream of American painting," but that "something . . . raises his work from a level of technical ornithological illustration to one of aesthetic quality."[5]

Although the relationship between art and science was closer than one might think during Audubon's life, few of Audubon's contemporaries understood that his careful depictions of birds and their environments in the early 1820s placed him in the forefront of the coming generation of American Romantic painters. They believed, with Emerson and Thoreau, that art and science were but different facets of the same edifice, and that nature, when carefully reproduced, held the keys to understanding the Creator's mystery. Audubon's thoroughgoing Romanticism would surprise no one who had studied his prints or read his essays. Even his personal extravagances and individualism marked him as a Romantic—before his paintings were known. For instance, he preferred secluding himself in the forest, observing and drawing his birds, to managing the family farm at Mill Grove or minding the store in Henderson and finally had only his art left as a possible living for his family. He was roaming the wildernesses of the Old Northwest and the Mississippi River valley a full decade before George Catlin embarked upon his Romantic quest of the exotic and noble savages of the upper Missouri River. The fact that Audubon, while in a destitute state, conceived of a giant publication that would encompass all the birds of America, reproduced life-size, costing more than $115,000 to produce, and selling for $1,000 per copy, was the perfect expression of a heroic ego in a heroic age.

His quest was genuine, however. Convinced that the birds of America had never been adequately cataloged or depicted, he was equally certain that he was the person for the job. Along with other naturalists of the era, Audubon felt that each time he discovered, illustrated, described, and shared knowledge of each new species of bird, he was helping piece together the "Great Chain of Being," link by link, just as Nuttall, Lesueur, Catlin, and the other explorer-artists were doing for the plants, animals, and aboriginal inhabitants of the American continent.

His inclinations are evident throughout his writings in the fantastic, descriptive, and sensual language of the Romantic, which he employed. His father inspired him to study birds, he explained in the first volume of the *Ornithological Biography*, "and to raise my mind towards their great Creator." On another occasion, he recalled that, "I marveled at Nature as the dawn presented her—in richest, purest array—before her Creator. . . .

Again I was full of desire to comprehend all I saw!" Ever alert to nature as he traveled throughout the country, he heard the "rumbling sound" of water and the "loud and strange noise" of Indian warfare. He saw the "rich and glowing hue" of the sun, "lofty hills," and "extensive plains," the "dismal clouds" and "awful phenomena" of a hurricane. He found a "wild and solitary spot" in the forest, and on the coast, he frequently enjoyed the "grandeur and beauty of those almost uninhabited shores."[6]

It is a cliché to say that Romantic art replaced the eighteenth century's objective view of the world with a subjective one, but as with most clichés, there is some truth in it. In Audubon's case, his paintings of birds appeared to be especially objective, particularly when one reads the accompanying text and realizes that, even in the most animated and violent images, he has depicted an important characteristic or defining moment of even so modest a bird as the tufted titmouse (Havell 39), curled over a small pine seed to crack it with his beak; or the twisted, stretching posture of the wood thrush (Havell 73) reaching for a berry. Audubon is, perhaps, at his best in the figures of the lesser terns (Havell 319), which, unlike most of his birds, lack the perspective that a habitat or landscape would have provided. Set against a dark void broken only by the clouds, the terns sail gracefully through the changing wind currents, a depiction that enhances their elegant form and flight. "Nothing can exceed the lightness of the flight of this bird," Audubon emphasized. "They move with swiftness at times, at others, balance themselves like hawks . . . then dart with . . . velocity." He suggests these movements with what might seem to some awkwardly posed birds, with their wings pointing up at sharp angles. But the angles are echoed in the clouds, which, as a result, seem to be a more inviting and natural setting for the terns.

Even so, Audubon anticipated criticism and wrote in 1831: "The positions may, perhaps, in some instances appear *outré*; but such supposed exaggerations can afford subject of criticism only to persons unacquainted with the feathered tribes; for, believe me, nothing can be more transient or varied than the attitudes or positions of birds."[7]

A close examination of the pictures, though, also clearly reveals Audubon's pervasive and subjective hand. A significant tenet of Romantic theory holds that the artificial barriers separating one part of nature from another should be broken down. A Romantic artist might accomplish this by painting a picture that is not complete within the frame; a detail of a larger image, perhaps—a landscape or a painting in which the main character might be gazing outside the frame—to suggest that some important element of the narrative is taking place outside the picture, in the larger world. Book illustrations became popular during the late eighteenth and early nineteenth centuries, in part because the development of wood engraving made it possible to print the images and the text at the same time and on the same page, with only one pass through the press. To the Romantics this meant that the barrier between text and illustration, between two art forms, had been cleared, and the resulting vignettes, which fade out at the edges to suggest that they are fragments of a larger world, are perfect expressions of Romantic sentiment. The vignettes are not complete, as might be suggested if they were framed or contained by a line. Thomas Bewick, whom Audubon greatly admired, gave the Romantic vignette its earliest and best expression, but Audubon's own small prints, especially the ones with the detailed habitats that fade out toward the edges, are equally good examples.[8]

Another aspect of that philosophy was based on the fundamental concept, as Emerson explained, that "behind nature, throughout nature, spirit is present." To him natural science broke down the barriers in nature, between people and animals, and showed the human to be a part of the "Great Chain of Being," a part of the universe, "with a ray of relation passing from him to every created thing." Emerson's contemporary Henry David Thoreau, perhaps of all the Romantic writers, felt the relationship applied especially to birds and observed, for example, that the great horned owl was a native of America long before the Anglos arrived, that the great blue heron "belongs to a different race from myself. . . . [but] I am glad to recognize him for a native of America,—why not an American citizen." Thoreau even tried to share the birds' perspective. "I found myself suddenly neighbour to the birds," he wrote of his experience at Walden, "not by having imprisoned one, but by having caged myself near them."[9]

Audubon painted much as Thoreau thought and wrote. He, too, broke down the barriers between humans and animals. He painted birds at eye level like earlier naturalists. But, unlike his predecessors, who painted them standing on the ground or perched on a branch, he painted them in genre scenes that invite the viewer to share the often moral stories or dilemmas depicted. What William Sidney Mount did for the Yankee, Charles Deas did for the frontiersman, and George Caleb Bingham did for the Missouri riverman, Audubon did for birds. Audubon's subscribers probably were not surprised to find themselves face to face with the wide-eyed and life-size wild turkey (Havell 1), startling and colorful as he may be, when they opened the first number of the double elephant folio, but they were surely shocked to find themselves confronting a ferocious peregrine falcon, in the process of devouring, along with his mate, the remains of a green-winged teal (Plate 60). In later plates, Audubon took viewers into the tangles in the jessamine vine, beside a rattlesnake as he attacked a family of mockingbirds, and hundreds of feet above the ground soaring with the golden eagle (Havell 181) and osprey (Havell 81).[10]

Audubon further broke down the barrier between people and nature by ascribing human expressions, feelings, and moral dramas to certain species. The female cormorant (Havell 266) personifies maternal love as she "gently caresses each [offspring] alternately with her bill." The blue jay (Plate 61) betrays his antisocial nature in one of Audubon's most beautiful compositions, a "rogue . . . sucking eggs which he has pilfered from the nest of some innocent Dove," and the brown thrashers (Havell 116) exhibit righteous indignation in defending their nest from an invading blacksnake. He described the peregrine falcons (O.P. 315, Havell 16, Bowen 20) as having "bloody rage at their beaks' ends, and . . . cruel delight in the glance of their daring eyes." Several other birds seem to have particularly violent and startled or angry expressions, such as the *Mocking Bird* (Havell 21), the *Virginian Partridge* (Havell 76), and the *Common Buzzard* (Havell 372).[11] These messages Audubon's viewers and readers accepted as faithful and accurate communication from nature itself, through the naturalist's writings and paintings.

Some critics have been so impressed by Audubon's views of atavistic and violent creatures that they have suggested these images were precursors of Darwin's theory of the survival of the fittest; more likely they were simply the result of close observation of nature over many years, seen through his Romantic lens. Others have criticized the unnatural attitudes of a number of his birds, such as the *Brasilian Caracara Eagle* (Havell 161) and the *Hudsonian Godwit* (Havell 258). They might not have considered that Audubon designed the poses for particular purposes. The back and front views of the caracara, for example, permit all the feathers to be seen, and the awkward spread of the godwit's wing allows Audubon to show the black color of the inner wing, thus distinguishing it from the black-tailed godwit of Europe.[12]

One of Audubon's best portraits is the bald eagle (Havell 11). Thinking that the immature specimen was a new species, he named it the "Bird of Washington" in honor of the first president of the nation and rendered a portrait as proud and complex as Gilbert Stuart or the Peales had ever painted. He explained:

> The name which I have chosen . . . may, by some, be considered as preposterous and unfit, but as it is indisputably the noblest bird of its genus that has yet been discovered in the United States, I trust I shall be allowed to honour it with the name of one yet nobler, who was the saviour of his country, and whose name

will ever be dear to it.... He had a nobility of mind, and a generosity of soul, such as are seldom possessed. He was brave, so is the Eagle; like it, too, he was the terror of his foes; and his fame, extending from pole to pole, resembles the majestic soarings of the mightiest of the feathered tribe. If America has reason to be proud of her Washington, so has she to be proud of her great Eagle.

Audubon had discovered his error by the time he wrote the *Ornithological Biography* and corrected the mistake, but the proud portrait remains, honoring the bird and the man.[13]

Although Audubon would have hated it, he is often compared with Catlin, whose Romantic portraits of Indians along the upper Missouri were rendered with a limited but colorful and dramatic palette. Both artists added to the scientific knowledge of the continent with lengthy expeditions into the wilderness that resulted in graphic and exotic portraits of unknown tribes. "We are proud of such men as Audubon and Catlin," James Hall wrote in Cincinnati's *Western Monthly Magazine*, "of native artists who are diffusing accurate knowledge of natural objects, in the land of their birth, by means of the elegant creations of the pencil." Catlin's *Four Bears, Second Chief, in Full Dress* (1832, oil on canvas, 29 by 24 inches, National Museum of American Art, Smithsonian Institution) is as proud as Audubon's bald eagle, but an equally revealing comparison is his masterful depiction of *Buffalo Bull's Back Fat, Head Chief, Blood Tribe* (Plate 63) with Audubon's *Ruffed Grouse* (Plate 64), among others. Both subjects display colorful and exotic decoration, finery from the heart of the wilderness for the Romantic to savor.[14]

The most famous examples of Romantic painting, however, are not genre or portrait painting but landscape. Romantic landscape painting grew out of a Rousseauean desire to glean spirituality from realistic depictions of nature—based on the belief that accurate depictions would reveal the greater truths of the Creator. The German artist Caspar David Friedrich may well have set the pace for Romantic landscapes in Europe; in America it was Thomas Cole. Employing the conventions of the picturesque and the sublime that had been well worked out in England and Europe, Cole reached a growing audience in America with canvases of native scenery studded with Romantic symbols, and he is credited with establishing America's first native artistic school, the Hudson River School. But Audubon had been painting idealized American landscapes as a part of the process of documenting birds for several years by the time Cole made his first sketching trip up the Hudson River. The *American Avocet* (Havell 318) and the

Spotted Sandpiper (Havell 310), both painted in 1821, contain superb examples of Louisiana landscape, and the tall coastal grass, in which the sharp-tailed sparrow has built his nest (Havell 149), is even reminiscent of Albrecht Dürer's *The Great Piece of Turf* (1503, watercolor and gouache on paper, Albertina, Vienna). But his best habitats and landscapes were obtained later, with the assistance of Joseph Mason and George Lehman: the *Black-Billed Cuckoo* (Havell 32), the *Ruffed Grouse* (Havell 41), the *Snowy Heron* (Havell 242), and the *Roseate Spoonbill* (Havell 321), among many others.

Audubon had planned for these painstaking and beautiful backgrounds to be a part of his bird paintings from the beginning, and they are a significant part of his Romantic as well as his scientific vision. He had taken young Mason down the Ohio and Mississippi rivers in 1820 and worked with the lad until he rendered extraordinary flowers, leaves, and branches for the habitats. The depiction of the large white flower and deep green leaves of the magnolia tree in the *Black-Billed Cuckoo* (Havell 32) is perhaps Mason's best, but the habitats in the *Yellow-Billed Cuckoo* (Havell 2), the *Purple Grackle* (Havell 24), and the *Mourning Dove* (Plate 65) are similarly impressive. This was not the first time subjects had been pictured in their natural habitat. Catesby had used habitats in his bird plates, Bartram in his images of American plants, with compositional elements to unite them with their environments. Lawson employed some simple landscapes behind some of Alexander Wilson's birds, and Charles Willson Peale exhibited his bird and animal specimens in their natural settings in his museum. However, the combination of habitats and birds is unlike those of any predecessor, and Audubon's intent is clear, not only from his paintings, but also from instructions that he passed along to his son shortly after he arrived in England. "Branches of Trees and Flowers I particularly wish him to do the size of nature and as closely as his talents will permit," he wrote Mrs. Audubon.[15]

Nor had any before Audubon composed such picturesque landscapes that cataloged the various regions of the new continent almost as thoroughly as he did its birds. He kept careful notes on Mason's and his own backgrounds and included information in the descriptions of his subjects. Upon his return to America in 1829, he employed George Lehman, a professional landscape painter, to furnish larger landscapes and cityscapes that are today among his most desirable prints: the *Long-Billed Curlew* (Havell 231) in Charleston harbor and the *Canvasback* (Havell 301) in Baltimore harbor, the *American Snipe* (Havell 243) and *Snowy Heron* (Havell 242) with views of South Carolina plantations in the background, and the *Louisiana Heron* (Plate 62) and the *Roseate Spoonbill* (Havell 321), with the marshes of the Florida Keys as the backdrop. So thoroughly schooled in their father's technique were Victor and John that when they began thinking of adding birds to the octavo edition in the early 1850s, it never occurred to them that they could improvise; they knew that it would take months to obtain the habitat information that would be required in the new illustrations. Where Thomas Cole's picturesque landscapes boosted American pride and awareness with sublime wildernesses and wild and exotic scenery, Audubon provided idealized but characteristic depictions of the countryside from the Atlantic Ocean to the Mississippi River.[16]

Just as Catlin's Indian portraits were "a literal and graphic delineation of... an interesting race of people, who are rapidly passing away from the face of the earth," Audubon might have intended some of his landscapes as memorials to scenes that he feared would vanish "before the encroachments of the white man." Recalling the Ohio River as he had seen it a decade before, when he had first navigated its waters, he wrote: "When I think of these times, and call back to my mind the grandeur and beauty of those almost uninhabited shores . . . when I remember that these extraordinary changes have all taken place in the short period of twenty years, I pause, wonder, and although I know all to be fact, can scarcely believe its reality." "Nature herself seems perishing," he wrote from Labrador in 1833.

"Whether these changes are for the better or the worse, I shall not pretend to say," Audubon continued,

> but . . . I feel with regret that there are on record no satisfactory accounts of the state of that portion of the country, from the time when our people first settled it. . . . However, it is not too late yet; and I sincerely hope that either or both [Cooper and Irving] . . . will ere long furnish the generations to come with those delightful descriptions which they are so well qualified to give, or the original state of a country that has been so rapidly forced to change her form and attire under the influence of increasing population.[17]

Not all of Audubon's landscapes were as carefully researched and depicted as these, of course. In two notable examples, the goshawk (Havell 141) and the black-throated guillemot (Havell 402), Audubon asked Havell to fill in the landscape backgrounds. The Arctic scene that Havell supplied for the guillemot is pleasing enough, although rather surreal, but Havell confused the perspective in the background for the goshawk, making one of the worst plates in the entire double elephant folio. Havell also provided backgrounds for the western duck (Havell 429), the white-legged oyster-catcher (Havell 427), and the plumed partridge (Havell 423), among others.

While these big and stunningly depicted birds with handsome landscapes in the background could be enjoyed purely for their picturesque qualities—qualities that are pleasing to the eye but that may, through a series of associations such as Audubon's, inspire one to loftier thoughts—his subscribers and admirers understood the deeper message. In Dr. George Parkman's copy of the *Ornithological Biography*, for example, near Audubon's description of a settler in the wilderness, "on his knees, with clasped hands, and face inclined upwards . . . ," a reader had noted a reference to William Cullen Bryant's poem, "A Forest Hymn," the first line of which begins, "The groves were God's first temples." George C. Shattuck, Sr., of Boston wrote Audubon to thank him for his "enterprise towards the manifestation of his [God's] creative powers," and his niece expressed her esteem for him as the "interpreter of nature." The editor of the *Saturday Courier* credited him with being a practitioner of the "elevating and soul-lifting Science of Natural History," while an *Albion* scribe concluded that, "Instruction in Natural History is here so happily blended with entertainment, that the perusal of its subject matter affords a delightful, and at the same time profitable pastime."[18]

Sensitive critics also immediately realized the impact of Audubon's work. French critic Philarète-Chasles asked his readers to "imagine a landscape wholly American, trees, flowers, grass, even the tints of the sky and the waters, quickened with a life that is real, peculiar, trans-Atlantic. On the twigs, branches, bits of shore, copied by the brush with the strictest fidelity, sport the feathered races of the New World, in the size of life, each in its particular attitude. . . . It is a real and palpable vision of the New World, with its atmosphere, its imposing vegetation, and its tribes which know not the yoke of man." Nevertheless, Charles Winterfield was one of the few American critics to see anything other than facts in his paintings:

> Shall we remind you that Audubon has elevated illustrative Ornithology from a state little short of a crude and unrecognized position as a feature,—along with "Cock Robin," and "Robinson Crusoe" epitomised—of the unmeaning toy-books of children, into the highest rank of Art which has striven truthfully to exhibit nature? Shall we remind you that in addition to having fixed it upon the profound basis of science as an illustrator, he has, as an accurate observer, carried its definition out of sight above predecessors or contemporaries, into the atmosphere of natural and practical philosophy—elaborating the delineations of sex, age, seasons and climate, into a precision and reality which must constitute the firm ground-work of future investigations?—in a word, that he has created,

60. Robert Havell, Jr., after John James Audubon, *Great-footed Hawk*, 1827. Hand-colored aquatint and engraving, 25 5/8 by 38 1/4 in. (plate). From *The Birds of America*, plate 16. Courtesy Stark Museum of Art, Orange, Texas. Ornithologists have criticized Audubon for depicting emotions that birds cannot express. These peregrine falcons, he said, have "bloody rage at their beaks' ends."

61. Robert Havell, Jr., after John James Audubon, *Blue Jay*, 1831. Hand-colored aquatint and engraving, 25 1/2 by 20 1/2 in. (plate). From *The Birds of America*, plate 102. Courtesy Stark Museum of Art, Orange, Texas. He described the blue jay as a "rogue...sucking eggs which he has pilfered from the nest of some innocent Dove."

62. Robert Havell, Jr., after John James Audubon, *Louisiana Heron*, 1834. Hand-colored aquatint and engraving, 20 3/4 by 25 7/8 in. (plate). From *The Birds of America*, plate 217. Courtesy Stark Museum of Art, Orange, Texas. The Florida Keys provided the setting for Lehman's landscape of an overgrown marsh accented with royal palms.

63. George Catlin, *Buffalo Bull's Back Fat, Head Chief, Blood Tribe*, 1832. Oil on canvas, 29 by 24 in. National Museum of American Art, Smithsonian Institution. Gift of Mrs. Joseph Harrison, Jr. Catlin set out to catalog and depict the various Indian tribes of North America just as Audubon was cataloging and documenting the birds.

64. Robert Havell, Jr., after John James Audubon, *Ruffed Grouse*, 1828. Hand-colored aquatint and engraving, 25 5/8 by 38 1/4 in. (plate). From *The Birds of America*, plate 41. Courtesy The National Gallery of Art, Washington, D.C. Gift of Mrs. Walter P. James. Audubon's ruffed grouse is every bit as exotic and colorful as Catlin's Indian portraits.

65. Robert Havell, Jr., after John James Audubon, *Mourning Dove*, 1827. Hand-colored aquatint and engraving, 26 3/4 by 20 3/4 in. (plate). From *The Birds of America*, plate 17. Courtesy Stark Museum of Art, Orange, Texas. Audubon had planned for accurate habitats to be a part of his paintings from the beginning. He employed young Mason to assist him with the detailed habitats and Lehman to provide the landscapes but painted these handsome camellias himself.

66. John James Audubon, *Common American Wild Cat*, 1845. Hand-colored lithograph by Bowen, 18 by 24 in. From *The Viviparous Quadrupeds of North America*, plate 1. Courtesy William S. Reese, New Haven, Connecticut. Audubon gave this print of the wild cat to Lieutenant Carleton, who assisted him on his western trip. Carleton gave Audubon a black bear skin and a set of elk horns in exchange.

67. Francis D'Avignon, *John James Audubon*, 1850. Lithograph, 11 1/16 by 9 9/16 in. (image). From the *Gallery of Illustrious Americans* (New York, 1850). Courtesy Stark Museum of Art, Orange, Texas. Inclusion in Mathew Brady's *Gallery of Illustrious Americans* certified Audubon's status as a genuine American hero of the Romantic period.

through Ornithology, the most alluring feature yet presented of that cheerful and broad philosophy which leads "through nature up to nature's God"? If you do not know all this, learn more of Audubon through his own works, and you will recognize it. We must defer to another No. a more familiar and pleasing intercourse with the man as well as naturalist, and with the wild natural scenes, which are the back-ground of his subjects.[19]

Audubon personified the American hunter-naturalist, heir to Daniel Boone and interpreter of America's natural paradise, for generations. After successfully presenting himself as the "American woodsman" in Europe, he continued the role at home. While in Philadelphia in November 1843, he "attracted general observation" on Chestnut Street, "clothed in a white blanket hunting coat and undressed otter skin cap; his 'beard was grizzled,' and, with his moustache, had been suffered to grow very long. On his shoulder, Natty-Bumpo-fashion, he carried his rifle, in a deer-skin cover." Through his writings and paintings, primarily the octavo edition of the *Birds*, many Americans vicariously shared his experiences in the wilderness. Congressman R. Barnwell Rhett, a subscriber who lived in Charleston, wrote Audubon in 1841 that "Mrs. Rhett takes great interest in your labours, and often describes to my little Boys, in glowing terms, taken from your works, the toils and the pleasure—the labours and the glory of being a great and enthusiastic Naturalist like Mr. Audubon." In *The Hunter-Naturalist*, author Charles W. Webber compared him to Boone. Both were "something of the Primitive Hunter and modern Field-Naturalist combined," he wrote. And Webber's might not have been the first such comparison. John G. Chapman's engraving of Boone, published in *Family Magazine* in 1836, was probably the inspiration for the handsome portrait that Audubon's sons painted of their father, seated on a rock and holding his rifle. His horse waits patiently at his right, while his hunting dog rests at his feet.[20]

The year before his death, Audubon received an even greater honor at the hands of the famous daguerreotypist Mathew Brady, who included him in his *Gallery of Illustrious Americans* (New York, 1850). Audubon was already an object of veneration, the focal point of much of the abstract admiration that Americans felt for Romantic values. Now lithographer Francis D'Avignon's copy of Brady's photograph portrait of Audubon (Plate 67) made his likeness widely available, a visible symbol of powerful and even more widely held beliefs.[21]

The Birds of America is an accomplishment whose stature grows year by year. The

Havell edition was simply one of the greatest books ever produced. Audubon reached his greatest audience, however, with the royal octavo edition, and it might well have been the medium by which the Romantic message reached its greatest public in America. "Audubon's Birds of America, and Audubon's Quadrupeds of America, will be great National Works, which will live to glorify his talents and perseverance so long as a love of Natural history shall endure," predicted the *Saturday Courier*. "It is folly to suppose that most of the people of the United States have not heard of this great contribution to the Literature, the Science and the Arts of America." While some of Audubon's peers recognized his talent and subscribed to the miniature edition, later artists specifically credited his aesthetic genius. A writer in the *New York Times* suggested that Winslow Homer's painting of *Wild Geese* was "worthy of Audubon," while Homer himself, one of the greatest of America's nineteenth- and early twentieth-century artists, paid tribute to the naturalist with *Right and Left* (1909, oil on canvas, National Gallery of Art), a dramatic composition of ducks that have just been shot, quite similar to Audubon's *Golden-Eye Duck* (Havell 342). Charles Birchfield, Marsden Hartley, and Ellsworth Kelly are only three among the many twentieth-century artists who were impressed with the more formal qualities of Audubon's work, which convinced Hartley that "a painting of life in the natural world may also be a work of art, and lose none of its scientific veracity."[22]

Audubon's science has long since been surpassed, but his form-shattering vision has had an enormous impact, and his dramatically rendered birds continue to resonate today. His was the romantic heart of American scientific observation, striving for objectivity but so freighted with moral beliefs and purposes that distortion unknowingly intervened. The eminent English critic Sacheverell Sitwell concluded that, "The 'American Woodsman' may be, within his limitations, the most considerable painter that the American continent, North and South, has yet produced." Yet even Sitwell probably did not realize how completely Audubon united unbridled artistic ability and the emerging American identity with nature during this formative period. Unlike the other Romantic artists, such as Allston, Cole, and Mount, who adapted classical myths and in some cases, derivative European artistic traditions to American situations and styles, Audubon brought out of the American wilderness, using American birds and their natural habitats, a completely new expression of American Romanticism. This expression came virtually at the last possible moment before natural history professionals turned toward more dispassionate and objective observation in their quest for the new science.[23]

Tables, Appendix, Notes, Bibliography, and Index

Table One

Printings of the First Royal Octavo Edition of *The Birds of America*

No.	Date	1st Pntg	Date	2nd Pntg	Date	3rd Pntg	Date	4th Pntg	Date	5th Pntg	Totals
1	12/16/39	300	1/18/40	300	3/12/40	325	6/6/40	200	1/9/41	350	1,475
2	12/28/39	300	1/11/40	200	3/14/40	325	6/6/40	258	1/9/41	326	1,409
3	1/11/40	210	2/10/40	290	3/26/40	250	6/6/40	260	1/9/41	335	1,345
4	1/31/40	300	2/10/40	200	3/25/40	250	6/6/40	250	1/9/41	339	1,339
5	2/20/40	500	3/26/40	250	6/6/40	250	1/9/41	250			1,250
6	3/7/40	500	3/25/40	250	6/20/40	250	1/9/41	250			1,250
7	4/4/40	750	7/18/40	250	1/9/41	250					1,250
8	4/18/40	750	7/18/40	250	1/9/41	250					1,250
9	5/8/40	750	7/18/40	250	1/9/41	320					1,320
10	5/23/40	1,000	1/30/41	313							1,313
11	6/13/40	1,000	3/13/41	311							1,311
12	6/27/40	1,000	3/13/41	310							1,310
13	7/11/40	1,000	3/13/41	310							1,310
14	8/1/40	1,000	3/13/41	310							1,310
15	8/8/40	1,000	4/3/41	298							1,298
16	9/10/40	1,000	4/3/41	292							1,292
17	9/19/40	1,000	4/3/41	289							1,289
18	10/2/40	1,000	4/3/41	286							1,286
19	10/10/40	1,000	4/3/41	272							1,272
20	10/24/40	1,000	4/3/41	251							1,251
21	11/14/40	1,000	4/24/41	250							1,250
22	11/28/40	1,000	4/24/41	252							1,252
23	12/19/40	1,000	4/24/41	250							1,250
24	1/2/41	1,250									1,250
25	1/30/41	1,250									1,250
26	3/13/41	1,250									1,250
27	4/10/41	1,250									1,250
28	6/8/41	1,500									1,500
29	6/8/41	1,500									1,500

30	6/8/41	1,500	1,500
31	5/20/41	1,250	1,250
32	5/20/41	1,250	1,250
33	6/5/41	1,250	1,250
34	6/14/41	1,250	1,250
35	6/19/41	1,250	1,250
36	6/21/41	1,250	1,250
37	6/27/41	1,250	1,250
38	8/13/41	1,250	1,250
39	8/28/41	1,250	1,250
40	9/12/41	1,250	1,250
41	9/23/41	1,250	1,250
42	10/9/41	1,250	1,250
43	10/27/41	1,250	1,250
44	11/15/41	1,250	1,250
45	11/27/41	1,250	1,250
46	1/10/41	1,250	1,250
47	12/24/41	1,250	1,250
48	1/22/42	1,250	1,250
49	2/5/42	1,250	1,250
50	2/19/42	1,250	1,250
51	3/26/42	1,150	1,150
52	4/2/42	1,150	1,150
53	4/9/42	1,150	1,150
54	4/22/42	1,150	1,150
55	5/14/42	1,150	1,150
56	5/25/42	1,150	1,150
57	6/11/42	1,050	1,050
58	6/25/42	1,050	1,050
59	7/6/42	1,050	1,050
60	7/18/42	1,050	1,050
61	7/30/42	1,050	1,050
62	9/1/42	1,050	1,050
63	9/17/42	1,050	1,050
64	10/8/42	1,050	1,050
65	10/18/42	1,050	1,050
66	10/29/42	1,050	1,050
67	?	1,050	1,050
68	?	1,050	1,050
69	12/16/42	1,050	1,050
70	12/27/42	1,050	1,050
71	1/18/43	1,050	1,050
72	2/1/42	1,050	1,050

Tables

73	2/20/43	1,050	1,050
74	3/11/43	1,050	1,050
75	3/13/43	1,050	1,050
76	4/8/43	1,050	1,050
77	4/27/43	1,050	1,050
78	4/27/43	1,050	1,050
79	5/19/43	1,050	1,050
80	6/2/43	1,050	1,050
81	7/15/43	1,050	1,050
82	7/18/43	1,050	1,050
83	7/20/43	1,050	1,050
84	8/12/43	1,050	1,050
85	11/20/43	1,050	1,050
86	11/20/43	1,050	1,050
87	11/20/43	1,050	1,050
88	11/20/43	1,050	1,050
89	11/20/43	1,050	1,050
90	11/20/43	1,050	1,050
91	2/7/44	1,050	1,050
92	2/7/44	1,050	1,050
93	2/7/44	1,050	1,050
94	2/7/44	1,050	1,050
95	2/7/44	1,050	1,050
96	5/29/44	1,050	1,050
97	5/29/44	1,050	1,050
98	5/29/44	1,050	1,050
99	5/29/44	1,050	1,050
100	5/29/44	1,050	1,050
Total Estimated Lithographs Printed			587,410

The information in this table comes from the "Birds of America Day Book" and the "Audubon Journals and Ledgers" at the Houghton Library, Harvard University.

Table Two

Publication of the Octavo Edition

Vol.	Numbers	Plates	Date
1	1–14	1–70	1839–1840
2	15–28	71–140	1840–1841
3	29–42	141–210	1841
4	43–56	211–280	1841–1842
5	57–70	281–350	1842
6	71–84	351–420	1843
7	85–100	421–500	1843–1844

This table, taken from Herrick, *Audubon the Naturalist* 2:404, has been corrected with information from the "Birds of America Day Book" in the Audubon Papers, Houghton Library, Harvard University, and Victor, note, New York, June 21, 1844, in Lucy Audubon to Audubon, n.d., box 1, folder 38, in Audubon Papers, Beinecke Library, Yale University.

Table Three

Subscribers to the Royal Octavo Edition of *The Birds of America* By State and City

State	Total	Percent	City	
Alabama	24	.02	Greensborough	2
			Mobile	22
Connecticut	12	.01	Hartford	12
District of Columbia	53	.04	Georgetown	4
			Washington	49
Delaware	2	.00	Wilmington	2
Florida	2	.00	St. Augustine	2
Georgia	20	.02	Augusta	3
			Forsythe	1
			Hopetown	1
			Macon	1
			Milledgeville	2
			Queensborough	1
			Savannah	11
Kentucky	7	.01	Lexington	4
			Louisville	2
			Owensborough	1
Louisiana	69	.06	Baton Rouge	1
			Jackson	1
			Lafayette City	1
			Lafourche	1
			New Orleans	61
			St. Francisville	1
			Vermilionville	1
Maryland	181	.15	Annapolis	11
			Baltimore	168
			Hagerstown	1
			Perrymansville	1
Massachusetts	362	.30	Boston	207

Tables

			Cambridge	6
			Charlestown	2
			Duxbury	2
			Fairhaven	1
			Hingham	2
			Kingston	1
			Lowell	24
			Nantucket	20
			New Bedford	49
			Northborough	1
			Oxford	1
			Pittsfield	3
			Plymouth	3
			Roxbury	1
			Salem	11
			Springfield	5
			Taunton	2
			Williams College	1
			Worcester	20
Michigan	1	.00	Detroit	1
Mississippi	6	.00	Grand Gulf	1
			Natchez	5
Missouri	2	.00	St. Louis	1
New Hampshire	4	.00	Dover	1
			Exeter	1
			Portsmouth	2
New Jersey	5	.00	Jersey City	1
			Moorestown	2
			Trenton	2
New York	176	.14	Albany	7
			Brooklyn	8
			Butternut	1
			Cohoes	1
			Cold Springs	1
			Long Island	1
			Morrisania	1
			New York City	142
			Ogdensburgh	1
			Rochester	1
			Sag Harbour	1
			Sing Sing	1
			Throgg's Point	1
			Troy	3

Tables

			Utica	2
			Waterloo	2
			West Point	1
Ohio	10	.01	Cincinnati	5
			Cleveland	1
			Columbus	1
			Steubenville	1
			Zenia	2
Pennsylvania	108	.09	Bedford	1
			Carlisle	1
			Cedar Hill	1
			Holmesburgh	1
			Lancaster	4
			Margaretta Fur	1
			New Bloomfield	1
			New Garden	1
			Penn Brook	1
			Philadelphia	74
			Pittsburgh	11
			Pittsfield	1
			Reading	9
			Uniontown	1
Rhode Island	17	.01	Providence	17
South Carolina	90	.07	Beaufort	5
			Charleston	69
			Columbia	8
			Courtsville	1
			Eddisto	1
			Georgeton	1
			Orangeburgh	2
			Waterborough	1
			Winnsburough	1
			X Roads	1
Virginia	52	.04	Norfolk	5
			Petersburgh	1
			Richmond	41
Wisconsin	1	.00	Fort Crawford	1
Canada	19	.02	Chippewa	1
			Cornwall	1
			Coteau de Lac	1
			Halifax	1
			Kingston	1
			Montreal	7

Tables

			Osnabruck	1
			Quebec	4
			Simcoe	1
			Toronto	1
China	2	.00	Macoa	2
Cuba	2	.00	Havana	1
			Matanzas	1
Great Britain	12	.01	London	7
			Welshpool	1
			York	1
Others				12
Totals	1,239	1.00		1,239

This table is prepared from the subscriber information that Audubon published in *The Birds of America*.

Table Four

Royal Octavo Edition Subscribers in New York and New Orleans
Who Subscribed to the Havell *Birds* or Folio *Quadrupeds*

	New York		New Orleans		Totals	
	No.	%	No.	%	No.	%
Havell	0	.00	0	.00	0	.00
Quadrupeds	18	.13	0	.00	18	.09

Table Five

Who Subscribed to the Royal Octavo *Birds* By Profession

	New York		New Orleans		Totals	
	No.	%	No.	%	No.	%
Accountant	0	.00	2	.03	2	.01
Artist	3	.02	0	.00	3	.01
Banker	6	.04	1	.02	7	.03
Bookkeeper	0	.00	1	.02	1	.00
Books	2	.01	0	.00	2	.01
Broker	3	.02	2	.03	5	.02
Business	2	.01	0	.00	2	.01
Construction	0	.00	1	.02	1	.00
Consul	2	.01	0	.00	2	.01
Doctor	6	.04	3	.05	9	.04
Editor	2	.01	0	.00	2	.01
Engineer	0	.00	1	.02	1	.00
Gentleman	1	.01	0	.00	1	.00
Government	2	.01	3	.05	5	.02
Importer	2	.01	0	.00	2	.01
Insurance	0	.00	1	.02	1	.00
Lawyer	17	.12	5	.08	22	.11
Library	3	.02	1	.02	4	.02
Manufacturer	1	.01	0	.00	1	.00
Merchant	32	.23	15	.25	47	.23
Military	2	.01	0	.00	2	.01
Notary Public	0	.00	1	.02	1	.00
Planter	0	.00	8	.13	8	.04
Preacher	1	.01	0	.00	1	.00
Taxidermist	1	.01	0	.00	1	.00
Unknown	54	.38	16	.26	70	.34
Totals	142	1.00	61	1.00	203	1.00

These tables were prepared, with the assistance of Florence M. Jumonville, Head Librarian, Historic New Orleans Collection, from city directories and other such publications, which are listed in the bibliography.

Appendix

1. John James Audubon, *The Birds of America, from Drawings Made in the United States and Their Territories.* 7 vols.

    Vol. 1. New York: J. J. Audubon; Philadelphia: J. B. Chevalier, 1840.
    Vol. 2. New York: J. J. Audubon; Philadelphia: J. B. Chevalier, 1840.
    Vol. 3. New York: J. J. Audubon; Philadelphia: J. B. Chevalier, 1841.
    Vol. 4. New York: J. J. Audubon; Philadelphia: J. B. Chevalier, 1842.
    Vol. 5. New York: J. J. Audubon; Philadelphia: J. B. Chevalier, 1843.
    Vol. 6. New York and Philadelphia: J. J. Audubon, 1843.
    Vol. 7. New York and Philadelphia: J. J. Audubon, 1844.

    This edition is probably the most common as well as the most desirable of the miniature editions. Various parts of this edition were printed many times, both text and plates, because Audubon ordered the printings as he needed them. The lithographic artists, R. Trembley, William E. Hitchcock, "A. V.," "J. C." (John Collins?), or "C. P." (Charles Parsons?), are identified on almost all of the prints. (Cited in a number of sources including Elliott Coues, *Birds of the Colorado Valley, a Repository of Scientific and Popular Information Concerning North American Ornithology*, 629–630; S. Dillon Ripley and Lynette L. Scribner, comps., *Ornithological Books in the Yale University Library, Including the Library of William Robertson Coe*, 13; John Todd Zimmer, *Catalogue of the Edward E. Ayer Ornithological Library*, 22; Anna H. Perrault, comp. and ed., *Nature Classics: A Catalogue of the E. A. McIlhenny Natural History Collection at Louisiana State University*, entry 893; and Lois Elmer Bannon and Taylor Clark, *Handbook of Audubon Prints*, 48, 50.)

2. John James Audubon, *The Birds of America, from Drawings Made in the United States and Their Territories.* 7 vols. New York: V. G. Audubon, 1856.

    A reissue of the 1840–1844 edition. Volume 1 contains a steel engraving of John James Audubon by H. B. Hall after a portrait by Henry Inman. The plates are quite similar to those of the first edition, except that tinted backgrounds have been added to most of them. They were made from the same stones with the exception of those that were broken or damaged or for some other reason remade. The legends on many of the plates were redone, and the lithographic artists' initials have vanished from many of the plates. Copies may be found in the Harry Ransom Humanities Research Center at the University of Texas at Austin and Cornell University among others. (Cited in Bannon and Clark, *Handbook of Audubon Prints*, 50; Coues, *Birds of the Colorado Valley*, 644.)

Appendix

3. John James Audubon, *The Birds of America, from Drawings Made in the United States and Their Territories.* 7 vols. New York: V. G. Audubon, Roe Lockwood and Son, 1859.

    A reissue of the 1856 edition. The imprint on the reverse of the title page reads: "Entered, etc., 1839," and "R. Craighead, Printer, Stereotyper, and Electrotyper, Caxton Building, 81, 83, and 85 Center Street." There are numerous differences in the plates, as noted in text. Copies are listed at Southern Methodist University, Yale University, and Louisiana State University. (Cited in Ripley and Scribner, comps., *Ornithological Books in the Yale University Library*, 13; Perrault, comp. and ed., *Nature Classics*, entry 894; Bannon and Clark, *Handbook of Audubon Prints*, 50; Zimmer, *Ayer Ornithological Library*, 22–23.)

4. John James Audubon, *The Birds of America, from Drawings Made in the United States and Their Territories.* 7 vols. New York: V. G. Audubon, Roe Lockwood and Son, 1860.

    This set, issued to accompany the Bien chromolithographs, often is found with no plates. Extant copies that do have illustrations probably were made up from the available plates resulting from the differing numbers that were printed. Copies are listed at the University of Minnesota Bio-Medical Library, the University of Michigan, the Harry Ransom Humanities Research Center at the University of Texas at Austin, and the Stark Museum of Art in Orange, Texas, among others. (Cited in Bannon and Clark, *Handbook of Audubon Prints*, 50; Zimmer, *Ayer Ornithological Library*, 23–24.)

5. John James Audubon, *The Birds of America, from Drawings Made in the United States and Their Territories.* 7 vols. New York: J. W. Audubon, Roe Lockwood and Son, 1861.

    A reissue of the 1859 edition. Cornell University and the Brooklyn Museum list copies; Louisiana State University, Yale University, Brown University, Texas A&M University, and Texas Southern University hold copies with no plates, which were probably intended to accompany the Bien chromolithographs. (Cited in Ripley and Scribner, comps., *Ornithological Books in the Yale University Library*, 13; Perrault, comp. and ed., *Nature Classics*, entry 895; Coues, *Birds of the Colorado Valley*, 661; Zimmer, *Ayer Ornithological Library*, 24–25; and Bannon and Clark, *Handbook of Audubon Prints*, 50.)

6. John James Audubon, *The Birds of North America: A Popular and Scientific Description of the Birds of the United States and Their Territories.* New York, 1863.

    Ghost edition, cited in Coues, *Birds of the Colorado Valley*, 666, of which no copies have been located.

7. John James Audubon, *The Birds of America, from Drawings Made in the United States and Their Territories.* 8 vols. New York: J. W. Audubon, Roe Lockwood and Son, 1865.

    A reissue of the 1859 edition with the pagination changed and bound in eight volumes instead of seven. (Coues, *Birds of the Colorado Valley*, 669.)

8. John James Audubon, *The Birds of America, from Drawings Made in the United States and Their Territories.* 8 vols. New York: George R. Lockwood, [1870].

    A reissue of the 1865 edition. Date is given as 1870 in the biography of Audubon, by George R. Lockwood, in volume 1. (Cited in Zimmer, *Ayer Ornithological Library*, 26–27.)

9. John James Audubon, *The Birds of America, from Drawings Made in the United States and Their Territories.* 8 vols. New York: George R. Lockwood, [1871].

    A reissue of the 1865 edition by George R. Lockwood of New York, the son of Roe Lockwood. Yale University and Cornell University list copies. (Cited in Ripley and Scribner, *Ornithological Books in the Yale University Library*, 13; Coues, *Birds of the Colorado Valley*, 686; Bannon and Clark, *Handbook of Audubon Prints*, 51.)

Appendix

10. John James Audubon, *The Birds of America, from Drawings Made in the United States and Their Territories.* 8 vols. New York: George R. Lockwood, 1889.
    Ohio State University and Colorado State University at Greeley list copies. (Cited in Christie's *Valuable Travel and Natural History Books*, lot 219.)
11. John James Audubon, *The Birds of America, from Drawings Made in the United States and Their Territories.* 7 vols. New York: Dover, 1967.
    Reprinted from the 1840–1844 edition, in octavo size, $5\frac{1}{2}$ by $8\frac{1}{2}$ in. Prints of birds are all in black and white. Bound in paper with an introduction by Dean Amadon in volume 1. (Cited in Bannon and Clark, *Handbook of Audubon Prints*, 51.)
12. John James Audubon, *The Birds of America, from Drawings Made in the United States and Their Territories.* 10 vols. Kent, Ohio: Volair Limited, 1979.
    Bound in 10 volumes. Volumes 1–7 contain the birds; volumes 8–10 contain the quadrupeds.
13. John James Audubon, *The Art of Audubon: The Complete Birds and Mammals: John James Audubon.* Introduction by Roger Tory Peterson. New York: Times Books, 1979.
    Color reproductions of the octavo plates from the first edition of both the *Birds* and the *Quadrupeds*, arranged by species. Text not included.
14. John James Audubon, *Audubon's Birds of America.* Text by George Dock, Jr. New York: Arrowood Press, 1987.
    Reproductions of thirty Havell plates as well as 470 of the third state, 1871 printing of the octavo bird plates. Some in color, most in black and white. The thirty plates not duplicated are those represented by the Havell plates.
15. John James Audubon, *Audubon's Birds of North America.* Introduction by Sheila Buff. Stamford, Conn.: Longmeadow Press, 1990.
    Color reproductions of the third state of the octavo plates. Text not included.
16. *Multimedia Birds of America: A Replica of the Complete Works of John James Audubon's* Birds of America *(1840–1844).* N.p.: CMC Research, Inc., 1990.
    Compact disk; contains all the text and color reproductions of all 500 of the first edition octavo plates. Also contains 150 birdcalls from the Cornell Library of Natural Sounds. The *Quadrupeds* are also available on CD-ROM.

Notes

CHAPTER ONE: THE "GREAT WORK"

1. Audubon referred to the octavo edition by several different names including "little work" and "little edition" in Audubon to John Bachman, Boston, December 8, 1839; "Small Edition," in Audubon to Maria Martin, Baltimore, February 29, 1840; both in bMS Am 1482, in John James Audubon Papers, Houghton Library, Harvard University, Cambridge, Mass. He referred to the "Small Work" in Audubon to S. G. Morton, Baltimore, February 26, 1840, in S. G. Morton Papers, American Philosophical Society, Philadelphia; and "*la petite Edition*" in Audubon to Victor G. Audubon, Charleston, November 4, 1833, in Howard Corning, ed., *Letters of John James Audubon, 1826–1840* 1:264. Audubon refers to the double elephant folio as the "Great work" in his letters to Lucy Audubon, Liverpool, May 16, 1827; and to Victor, New York, September 15, 1833, and Richmond, April 15, 1840, in Corning, ed., *Letters* 1:23, 248; 2:258. See also Audubon to S. G. Morton, New York, September 9, 1839, in Morton Papers; and David M. Lank, "Natural History Illustration: An Introductory Essay," in Anna H. Perrault, comp. and ed., *Nature Classics: A Catalogue of the E. A. McIlhenny Natural History Collection at Louisiana State University*, 53.

2. Christine E. Jackson, *Bird Etchings: The Illustrators and Their Books, 1655–1855*, 243–246; London *Monthly Chronicle* (September 1840), quoted in *Albion* (New York), September 26, 1840, p. 318, col. 1. Collector H. Bradley Martin's set of the double elephant folio sold at Sotheby's in New York on June 6, 1989, for $3,960,000; the University of Edinburgh set brought $4,070,000 in the spring of 1992. Out of the approximately 176 complete copies printed, there are, perhaps, fewer than 125 remaining with no more than 14 remaining in private hands. All 435 prints in the double elephant folio have been reproduced on numerous occasions: in book format by the Macmillan Company in 1937 and Abbeville Press in 1985; and in full-size prints by the Johnson Reprint Corporation in 1973 (unbound, $5,940; bound, $6,960), and Abbeville in 1985 (bound, $12,500). Waldemar H. Fries, *The Double Elephant Folio: The Story of Audubon's Birds of America*, 225–350, 369, 373.

3. Charles Sellers, *The Market Revolution: Jacksonian America, 1815–1846*, 364–395; *Saturday Courier* (Philadelphia), September 9, 1843, p. 2, col. 5. The Martin copy of the royal octavo edition brought $18,700 at the same auction, mentioned in note 2. Audubon talked about the small edition years before he actually began work on it. See Audubon to Victor Audubon, Charleston, November 4, 1833, in Corning, ed., *Letters* 1:264. See Appendix for the various editions. Some of the twentieth-century

reprints of the plates include Sheila Buff, ed., *Audubon's Birds of North America*; George Dock, Jr., ed., *Audubon's Birds of America*; and Roger Tory Peterson, ed., *The Art of Audubon: The Complete Birds and Mammals: John James Audubon*. The compact disk version of both the *Birds*, with 150 birdcalls, from the Cornell Library of Natural Sounds, and the *Quadrupeds* is available from, among others, DAK Industries in Canoga Park, California. See DAK catalogue (Winter 1992), 56–57. The price for the Audubon disks, plus two other disks, is presently $129.

4. Earlier biographers such as Francis Hobart Herrick, *Audubon the Naturalist: A History of His Life and Time* 1:174; Alexander B. Adams, *John James Audubon: A Biography*, 28; and Constance Rourke, *Audubon*, 12, 24, 193, take Audubon at his word about studying with David. But later scholars, most notably Alice Ford and Michael Harwood, have raised questions. It is claimed that Charles Lucien Bonaparte asked David about Audubon, and that David is said not to have remembered him, but because the information comes from Audubon's enemies, who spent years attacking him, this evidence may be questioned. See Alice Ford, *John James Audubon: A Biography*, 22, 42, 45–74; Alice Ford, ed., *The 1826 Journal of John James Audubon*, 170 and note 32, 65–66, note 15, and 302–303, note 71; Michael Harwood, "A Watershed for Ornithology," in Alton A. Lindsey, *The Bicentennial of John James Audubon*, 31–47; Mary Durant and Michael Harwood, *On the Road with John James Audubon*, 291; and Donald A. Shelley, "Audubon's Technique As Shown in His Drawings of Birds," *Antiques* 49 (1946): 354–357. Edward H. Dwight, *Audubon Watercolors and Drawings*, 6, suggests that David's influence might have been most evident in Audubon's charcoal portraits. For a summary of this argument, see Gloria K. Fiero, "Audubon the Artist," in James H. Dormon, ed., *Audubon: A Retrospective*, 34–60. A study of Audubon's stylistic abilities at the time he arrived in America suggests that he was self-taught.

5. Ford, *Audubon: A Biography*, 61–64.

6. The story of Audubon's meeting with Alexander Wilson has been told many times. For example, see Herrick, *Audubon the Naturalist* 1:220–226; Ford, *Audubon: A Biography*, 77–78; L. Clark Keating, *Audubon, the Kentucky Years*, 6–49; Robert Henry Welker, *Birds and Men: American Birds in Science, Art, Literature, and Conservation, 1800–1900*, 48–58; and Daniel Francis McGrath, "American Colorplate Books, 1800–1900," Ph.D. diss., 9. Elsa Guerdrum Allen, "The History of American Ornithology before Audubon," *Transactions of the American Philosophical Society* 41 (1951): 558–564; and Robert Cantwell, *Alexander Wilson, Naturalist and Pioneer*, 159, 192, 200–202, tell of it from Wilson's point of view.

7. Allen, "History of American Ornithology," 562–563; Ford, *Audubon: A Biography*, 101–109; Carolyn E. Delatte, *Lucy Audubon, a Biography*, 38–99; Maria R. Audubon, ed., *Audubon and His Journals* 1: 37, 38.

8. Ford, *Audubon: A Biography*, 101–115; Delatte, *Lucy Audubon*, 103–107; Howard Corning, ed., *Journal of John James Audubon, Made during His Trip to New Orleans in 1820–1821*, 3.

9. John James Audubon, *My Style of Drawing Birds*, 16–17, 22. Audubon also published a similar essay in *Edinburgh Journal of Science* in 1828, and Maria Audubon, ed., included a version of it in *Audubon and His Journals* 2:524–525. See also Carlotta J. Owens, *John James Audubon*, [19–20]; Edward H. Dwight, *Audubon Watercolors and Drawings*, 15, 28. If Audubon had studied with David, he probably would have been taught—or at least would have observed—a similar technique, with the horizontal/vertical grid in front of the subject. Fiero, "Audubon the Artist," 47–48.

10. Ford, *Audubon: A Biography*, 121–122, 133–134, 152, 156; Corning, ed., *Journal of John James Audubon*, 126–127, 156–157. Also, Adam Gopnik, "Audubon's Passion," *New Yorker* 67 (February 25, 1991): 96; Gary A. Reynolds, *John James Audubon and His Sons*, 27–28. Kathryn Hale Proby, "Audubon in Lou-

isiana," in Dormon, ed., *Audubon: A Retrospective*, 21–33, esp. 27–29. Durant and Harwood, *On the Road*, 272, speculate that Mrs. André was one of Bernard Xavier Philippe de Marigny de Mandeville's mistresses. Audubon also met the artist John Vanderlyn, who noticed Audubon's physical resemblance to Andrew Jackson and asked him to pose for the body of the full-length portrait of Jackson that he was painting. Audubon is featured as the orderly in the background of the finished painting. See Ford, *John James Audubon*, opposite page 146, for a reproduction of the painting.

11. Dwight, *Audubon Watercolors*, 5, 28; Audubon, *My Style*, 19, 22; Donald C. Peattie, *Green Laurels: The Lives and Achievements of the Great Naturalists*, 229; Marshall B. Davidson, ed., Introduction to *Original Water-Color Paintings by John James Audubon for the Birds of America* 1: plate 55; Charles Coleman Sellers, *Mr. Peale's Museum: Charles Willson Peale and the First Popular Museum of Natural Science and Art*, 19; Edward P. Alexander, *Museum Masters: Their Museums and Their Influence*, 61; Allen, "History of American Ornithology," 465.

12. Michael Harwood, *Audubon Demythologized*, 10; Ford, *Audubon: A Biography*, 137–138.

13. Allen, "History of American Ornithology," 402–404, 411, 426–442, 463–478.

14. Welker, *Birds and Men*, 77; Sacheverell Sitwell, Handasyde Buchanan, and James Fisher, *Fine Bird Books, 1700–1900*, 9; Georgia B. Barnhill, "The Publication of Illustrated Natural Histories in Philadelphia, 1800–1850," in Gerald W. R. Ward, ed., *The American Illustrated Book in the Nineteenth Century*, 53–54; C. Jackson, *Bird Etchings*, 35–37; Lynn L. Merrill, *The Romance of Victorian Natural History*, 80–81; Lynn Barber, *The Heyday of Natural History, 1820–1870*, 14–15, 86–98; and Margaret Curzon Welch, "John James Audubon and His American Audience: Art, Science, and Nature, 1830–1860," Ph.D. diss., 20–21; Alan Feduccia, ed., *Catesby's Birds of Colonial America*, 4–5.

15. William H. Goetzmann and William N. Goetzmann, *The West of the Imagination*, 3–14; Welker, *Birds and Men*, 62–63. Brian W. Dippie, *Catlin and His Contemporaries: The Politics of Patronage*, 157–263, discusses some of the most important government-sponsored publications, including Schoolcraft's *Historical and Statistical Information Respecting the History, Condition and Prospects of the Indian Tribes of the United States* and Squier's *Ancient Monuments of the Mississippi Valley, Comprising the Results of Extensive Original Surveys and Explorations*, which were published about the same time as the extensive government surveys such as the Wilkes expedition, the Mexican boundary survey, and the Pacific Railroad surveys.

16. Lucy Audubon, ed., *The Life of John James Audubon, the Naturalist*, 100–105; Ford, *Audubon: A Biography*, 147–153; Welker, *Birds and Men*, 63; Philadelphia *Mercury*, quoted in the *Saturday Courier* (Philadelphia), November 11, 1843, p. 2, col. 2; Audubon to Lucy Audubon, Baltimore, March 1, 1840, in Corning, ed., *Letters* 2:239. Audubon even mentioned studying under David in his journal, probably because he frequently let the Rathbone ladies read it. Ford, ed., *1826 Journal*, 170.

17. William Dunlap, *History of the Rise and Progress of the Arts of Design in the United States* 3:204–205; L. Audubon, ed., *Life of John James Audubon*, 101; Ford, *Audubon: A Biography*, 147–152, 162–168, 361.

18. Dunlap, *Rise and Progress* 3:204–205; Ford, *Audubon: A Biography*, 148–150, 165–168, 360–361. It has often been said that the *Boat-Tailed Grackles* in Bonaparte's *American Ornithology; or, The Natural History of Birds Inhabiting the United States, Not Given by Wilson*, plate 4, was Audubon's first published work, but Ford, *Audubon: A Biography*, 162–172, 360–361, discovered a letter from George Ord to Charles Waterton explaining that the picture had been completely redrawn by Alexander Rider. The drawing in the collection of the Academy of Natural Sciences of Philadelphia, which bears both Rider's and Audubon's names is probably Rider's work after Audubon. The difference between Audubon's

work and the drawing, proved by the subsequent discovery of Audubon's original work, substantiates the claim. See Scott Gentling, John Graves, and Stuart Gentling, *Of Birds and Texas*, 87–90.

19. Winterfield, "American Ornithology," *American Review: A Whig Journal* 1 (March 1845): 267.

20. C. Jackson, *Bird Etchings*, 22. Frederic Remington was still complaining about engravers as late as the 1880s. See Estelle Jussim, *Frederic Remington, the Camera and the Old West*, 16.

21. L. Audubon ed., *Life of John James Audubon*, 106–107; Ford, *Audubon: A Biography*, 146–147, 152–173.

22. John Wilmerding, *American Art*, 36–44, 47–69; Cantwell, *Alexander Wilson*, 125, 137–138, 141.

23. Dwight, *Audubon Watercolors*, 28.

24. John P. Bakewell to Euphemia Gifford, New Orleans, May 13, 1826, in the Audubon Papers, Stark Museum of Art, Orange, Texas; Ford, ed., *1826 Journal*, 45, 51, 56, 204, and 379 (Nolte to Rathbone, New Orleans, May 26, 1826). Euphemia Gifford was the daughter of Elizabeth Woodhouse, great-aunt of Lucy Audubon, and Richard Gifford, Vicar of Duffield.

25. Ann Gordon to Euphemia Gifford, Liverpool, August 5, 1826, in Audubon Papers, Stark Museum of Art; Ford, ed., *1826 Journal*, 72.

26. Herrick, *Audubon the Naturalist* 1:367, quoting Sir Walter Scott. Audubon to Lucy Audubon about his "unshorn tresses," in Corning, ed., *Letters* 1:4. The self-portrait is illustrated in Dwight, *Audubon Watercolors*, [2].

27. Fries, *Double Elephant Folio*, 5, 8, 9–10; M. Audubon, ed., *Audubon and His Journals* 1:103, 135; Ford, ed., *1826 Journal*, 175–176, 244, and 366–367 (Audubon to Thomas S. Traill, Edinburgh, October 28, 1826); Audubon to Victor Audubon, Liverpool, September 1, 1826, in Corning, ed., *Letters* 1:3–4. The late bookseller Warren Howell of San Francisco would have agreed with Bohn and frequently told his employees that large picture books were difficult to sell because people wanted to be able to look at them. Barber, *Heyday of Natural History*, 92, 253; Jeannette E. Graustein, *Thomas Nuttall, Naturalist: Explorations in America, 1808–1841*, 164; Welch, "Audubon and His American Audience," 109–110; C. Jackson, *Bird Etchings*, 238. The Wernerian Society of Natural History at the University of Edinburgh was founded in 1808 and named after Abraham G. Werner, who was an honorary member of the society himself. See Archibald Geikie, *The Founders of Geology*, 327.

28. M. Audubon, ed., *Audubon and His Journals* 1:117–118, 258; Audubon to Lucy Audubon, Manchester, September 20, 1827, in Corning, ed., *Letters* 1:39–40; Fries, *Double Elephant Folio*, 7, 137; Ford, *Audubon: A Bibliography*, 229; Christine E. Jackson, *Bird Illustrators: Some Artists in Early Lithography*, 14–15; S. Peter Dance, *The Art of Natural History: Animal Illustrators and Their Work*, 94–95; Welch, "Audubon and His American Audience," 101–109.

29. Ford, *Audubon: A Biography*, 198–202; Fries, *Double Elephant Folio*, 8–10, 385–389, reproduces the prospectus. It is photographically reproduced between pages 436 and 437. See also Audubon to Lucy Audubon, Edinburgh, December 21, 1826, in Corning, ed., *Letters* 1:8. The $10 price for one fascicle in 1826 would be the equivalent of approximately $137 today. See John J. McCusker, "How Much Is That in Real Money? A Historical Price Index for Use As a Deflator of Money Values in the Economy of the United States," *Proceedings of the American Antiquarian Society, a Journal of American History and Culture Through 1876* 101 (Pt. 2, 1992), 327, 332.

30. Audubon to Lucy Audubon, Liverpool, May 16, 1827, in Corning, ed., *Letters* 1:23–24, 26; Herrick, *Audubon the Naturalist* 1:358; M. Audubon, ed., *Audubon and His Journals* 1:156.

31. Ford, ed., *1826 Journal*, 292–293; M. Audubon, ed., *Audubon and His Journals* 1:211; Ford, *Audubon:*

A Biography, 219; Herrick, *Audubon the Naturalist* 1:359 (quote from the French critic), 367 (quote from Scott).

32. Ford, ed., *1826 Journal*, 262; Ford, *Audubon: A Biography*, 203–204.

33. M. Audubon, ed., *Audubon and His Journals* 1:221; Audubon to Lucy Audubon, Liverpool, May 16, 1827, and Manchester, September 20, 1827, in Corning, ed., *Letters* 1:27, 39–40; Ford, *Audubon: A Biography*, 221–225; Fries, *Double Elephant Folio*, 19, 198.

34. Audubon to Lucy Audubon, London, August 6, 1827, in Corning, ed., *Letters* 1:29–34. See also Ford, *Audubon: A Biography*, 225–226; Fries, *Double Elephant Folio*, 399; C. Jackson, *Bird Etchings*, 236–237.

35. George Alfred Williams, "Robert Havell, Junior, Engraver of Audubon's 'The Birds of America,'" *Print Collector's Quarterly* 7 (October 1917): 234–240; C. Jackson, *Bird Etchings*, 236.

36. C. Jackson, *Bird Etchings*, 236; Audubon to Lucy Audubon, London, August 6, 1827, and Edinburgh, November 12, 1827, in Corning, ed., *Letters* 1:29–30, 42; Fries, *Double Elephant Folio*, 23–26; Ford, *Audubon: A Biography*, 226. Havell surely had to have the help of able assistants in engraving 425 plates in eleven years. He would have had to average almost 39 plates per year, an impossible job for only one person.

37. Fries, *Double Elephant Folio*, 211–212, 215–224. The Amon Carter Museum in Fort Worth, Texas, has the first three numbers of *Birds of America*, which Audubon shipped, probably in September or October 1827, to Bonaparte in America, not realizing that he had left America for Italy. The prints languished in the customs house until they were retrieved by Bonaparte's friend William Cooper. They remained in Cooper's family until his grandson, Hermann F. Cuntz, sold them to Rockwell Gardiner, a Connecticut book and antique dealer, who sold them to Harry Shaw Newman of the Old Print Shop in New York. Newman then sold them to the Amon Carter Museum. William S. Reese, "The Bonaparte Audubons at the Amon Carter Museum and the Friendship of John James Audubon and Charles Lucien Bonaparte," in Ron Tyler, ed., *Prints of the American West: Papers Presented at the Ninth Annual North American Print Conference*, 13–24.

38. C. Jackson, *Bird Etchings*, 29; Suzanne M. Low, *An Index and Guide to Audubon's* Birds of America: *A Study of the Double Elephant Folio of John James Audubon's* Birds of America, *as Engraved by William H. Lizars and Robert Havell*, 21–24.

39. Audubon to Lucy Audubon, Liverpool, December 26, 1827, in Corning, ed., *Letters* 1:55.

40. Herrick, *Audubon the Naturalist* 1:1, 410–412. The second quote is also in Ford, *Audubon: A Biography*, 243; then, on page 487, quoting *Le Moniteur Universal* (Paris), October 1, 1828: "The most magnificent monument which has yet been raised to ornithology."

41. Audubon to Victor Audubon, London, January 20, 1829; Audubon to Lucy Audubon, London, February 1, 1829, and New York, May 10, 1829; all in Corning, ed., *Letters* 1:79–81; Fries, *Double Elephant Folio*, 37, 45; Ford, *Audubon: A Biography*, 255–256, 260.

42. Ford, *Audubon: A Biography*, 259–260; Bruce Sinclair, *Philadelphia's Philosopher Mechanics: A History of the Franklin Institute, 1824–1865*, 199; Audubon to Lucy Audubon, New York, May 10, 1829, in Corning, ed., *Letters* 1:85.

43. Fries, *Double Elephant Folio*, 38; Ford, *Audubon: A Biography*, 260–261; Dwight, *Audubon Watercolors*, 40. Audubon promised in the prospectus to include eggs in the double elephant folio, but he probably did not do so because he ran out of space. He had so many birds to add at the last minute that

he increased the number of plates from 400 to 435; still, there was no room for eggs. Fries, *Double Elephant Folio*, 95–96.

44. Audubon to Lucy Audubon, New York, May 10, 1829; Audubon to Victor Audubon, Philadelphia, July 18, 1829, and Great Pine Swamp, Northampton County, Pennsylvania, August 25, 1829; all in Corning, ed., *Letters* 1:81–86, 88–92, 93–98. Also, Ford, *Audubon: A Biography*, 261–266; Delatte, *Lucy Audubon*, 194–212; Durant and Harwood, *On the Road*, 297–298.

45. Ford, *Audubon: A Biography*, 267–270; Fries, *Double Elephant Folio*, 39–40; Audubon to Robert Havell, October 24 and 27, 1829, in Corning, ed., *Letters* 1:98–100, 100–102; Audubon to Havell, Beech Grove, Louisiana, December 16, 1829, in Herrick, *Audubon the Naturalist* 1:434.

46. Audubon to Robert Havell, Liverpool, June 7, 1830, Birmingham, June 29, 1830, New Castle, September 30, 1830, and Edinburgh, November 30, 1830, in Corning, ed., *Letters* 1:107, 112, 118, 123; Ford, *Audubon: A Biography*, 274–275; Fries, *Double Elephant Folio*, 42–43. Fries, who looked at thousands of different prints as he prepared his book, commented that "Audubon was unduly meticulous and that he found grave faults where they in fact existed only in small degree. It occasionally has been possible for this researcher simultaneously to compare prints struck off from the same plate. On the whole, the variations in colors have been slight. The differences have been noticeable principally in the shades of the blues and the greens; the former color would be too dark, a criticism often made by the naturalist, while in the latter there would be a slight yellow tinge. On the whole, however, the work of the colorists has shown a remarkable uniformity." See pages 43, 46. Audubon had no more complaints after Robert, Sr., retired and Robert, Jr., took over the coloring as well. Williams, "Robert Havell, Junior," 248.

47. Herrick, *Audubon the Naturalist* 1:438; Fries, *Double Elephant Folio*, 47–55.

48. Audubon to Victor Audubon, Edinburgh, February 21, 1831; Audubon to Dr. McMurtry, Edinburgh, February 21, 1831; and Audubon to Robert Havell, Edinburgh, March 12 and 23, 1831; all in Corning, ed., *Letters* 1:126, 127–130, 131, 132–133. Also, Fries, *Double Elephant Folio*, 50–51, 140; C. Jackson, *Bird Etchings*, 204, 206. Audubon did not know that all he had to do to secure an American copyright was deposit a copy of the title page of the *Ornithological Biography* in the office of the district court. See Edward Everett to Audubon, Charlestown, Massachusetts, May 19, 1831, in Herrick, *Audubon the Naturalist* 1:449.

49. Fries, *Double Elephant Folio*, 18–19, 44, 54; Audubon to Henslow, London, April 28, 1831, in John James Audubon, *Three Letters of John James Audubon to John Stevens Henslow*, [5].

50. Audubon to Lucy Audubon, Charleston, November 7, 1831, in Corning, ed., *Letters* 1:148; Ford, *Audubon: A Biography*, 287, 289; James B. Bell, *John James Audubon: A Selection of Watercolors in The New-York Historical Society*, 22, 25–26; M. Audubon, ed., *Audubon and His Journals* 1:[560].

51. Audubon to Lucy Audubon, Charleston, October 23, 1831, in Corning, ed., *Letters* 1:143; Ford, *Audubon: A Biography*, 283, 285; Fries, *Double Elephant Folio*, 152, 158–159.

52. Kathryn Hale Proby, *Audubon in Florida, with Selections from the Writings of John James Audubon*, 14–49; Audubon to John Bachman, Philadelphia, July 1, 1832; and Audubon to Victor Audubon, Boston, February 24, 1833; both in Corning, ed., *Letters* 1:195, 200–201. Also, Audubon to Victor, Boston, February [March] 5, 1833, in Herrick, *Audubon the Naturalist* 2:35–36, and see also p. 26n; Ford, *Audubon: A Biography*, 303, 358; Audubon to family, Boston, February 5, 1833, in Ruthven Deane, "A Hitherto Unpublished Letter of John James Audubon," *Auk* 22 (April 1905): 171. For the *Marsh Hens*, see the *National Gazette and Literary Register* (Philadelphia), September 13, 1832, and Richard S. Field, "Audubon's Lithograph of the Clapper Rail," *Art Quarterly* 29 (Spring 1966): 147–153.

53. Ford, *Audubon: A Biography*, 307–310, 358.

54. Hans Huth, *Nature and the American: Three Centuries of Changing Attitudes*, 14–25, 30–36, 44–47; Dunlap, *Rise and Progress* 2:402–408; Ford, *Audubon: A Biography*, 319–320.

55. Audubon to John Bachman, Philadelphia, July 1, 1832, in Corning, ed., *Letters* 1:195; Bayard Tuckerman, ed., *The Diary of Philip Hone, 1828–1851* 1:73; Ford, *Audubon: A Biography*, 291–310, 323–342.

CHAPTER TWO:
COMPLETION OF THE DOUBLE ELEPHANT FOLIO

1. Audubon to Victor Audubon, New York, September 15, 1833, in Corning, ed., *Letters* 1:247–249.

2. Herrick, *Audubon the Naturalist* 1:443–445; and Ford, *Audubon: A Biography*, 318.

3. Audubon to Victor Audubon, New York, September 15, 1833, in Corning, ed., *Letters* 1:247–249.

4. Audubon to Victor Audubon, New York, September 23, 1833, and Charleston, November 4, 1833, in Corning, ed., *Letters* 1:259, 264–265.

5. Audubon to Victor Audubon, Charleston, November 24 and December 21, 1833, in Corning, ed., *Letters* 1:269, 276.

6. Ford, *Audubon: A Biography*, 323; Audubon to John Stevens Henslow, London, January 30, 1835, in Audubon, *Three Letters*, [7]; Reynolds, *Audubon and His Sons*, 51.

7. Welch, "Audubon and His American Audience," 74, 96; Ford, *Audubon: A Biography*, 328, 336.

8. Audubon to John Bachman, London, June 13, 1836, New York, October 2, 1836, and Philadelphia, October 23, 1836, in Corning, ed., *Letters* 2:122, 133–134, 135–136; Fries, *Double Elephant Folio*, 98–99; Graustein, *Thomas Nuttall*, 277–319, and especially see 318–319.

9. Audubon to [?], Charleston, February 13, 1837, in Corning, ed., *Letters* 2:139–140; Mary Bell Hart, *Audubon's Texas Quadrupeds: A Portfolio of Color Prints*, 29. For information on the Long expedition, see Edwin James, comp., *Account of an Expedition from Pittsburgh to the Rocky Mountains, under the Command of Major Stephen H. Long*, 387; for information on Mier y Terán and Berlandier, see Ohland Morton, *Terán and Texas: A Chapter in Texas-Mexican Relations;* Jean Luis Berlandier, *Journey to Mexico during the Years 1826 to 1834;* and John C. Ewers, ed., *The Indians of Texas in 1830 by Jean Louis Berlandier*. The only nineteenth-century publication to result from Berlandier's work is Jean Luis Berlandier and Rafael Chovel, *Diario de viage de la Comisión de Limites*. Drummond's Texas letters are published in the *Companion to the Botanical Magazine* (London) 1:39–44, and see especially 39–41. Samuel Wood Geiser provides additional information on Drummond in "Thomas Drummond," *Southwest Review* 25 (Summer 1930): 478–512, as does Mary Austin Holley in *Texas*, vii–viii; the quote is found on page 100. Drummond worked under the patronage of Sir William Jackson Hooker, who was then Regius Professor of Botany at the University of Glasgow and would later become Keeper of the Royal Botanical Gardens at Kew, outside London.

10. Fries, *Double Elephant Folio*, 100–103, 152; Ford, *Audubon: A Biography*, 346–348; Audubon to his friend Edward Harris, New Orleans, March 4, 1837, pfMS Am 21, in Audubon Papers, Houghton Library, and Robert Buchanan, ed., *The Life and Adventures of John James Audubon, the Naturalist, Edited by Robert Buchanan from Materials Supplied by His Widow*, 329–331, which contains extensive passages

from Audubon's Texas journal. Samuel Wood Geiser, "Naturalist of the Frontier: Audubon in Texas," *Southwest Review* 26 (Autumn 1930): 109–135, used Buchanan's work, plus passages from Audubon, *Ornithological Biography; or An Account of the Habits of the Birds of the United States of America*, and *The Birds of America, from Drawings Made in the United States and Their Territories*, to reconstruct Audubon's Texas journals, which, along with John H. Jenkins, *Audubon and Texas*, summarizes Audubon's experience in Texas. Samuel Wood Geiser's essay is included in his *Naturalists on the Frontier*, 2d ed., without notes. See also Stanley Siegel, *A Political History of the Texas Republic, 1836–1845*, 56–58; Joe B. Frantz, *Gail Borden: Dairyman to a Nation*, 123–126; Jim Dan Hill, *The Texas Navy: In Forgotten Battles and Shirtsleeve Diplomacy*, 68–73; see Buchanan, *Life and Adventures of Audubon*, 343, for the quote about the secret mission.

11. *Telegraph and Texas Register* (Houston), May 2, 1837, p. 2, col. 2. See also Frantz, *Gail Borden*, 125; and Buchanan, *Life and Adventures of Audubon*, 340–341. Unfortunately, John's sketch of Galveston is not known to exist. The earliest published picture of Galveston may be the frontispiece in Matilda Houstoun, *Texas and the Gulf of Mexico; or Yachting in the New World*, vol. 2. Charles Hooton painted the village in 1839 (oil on canvas, Rosenberg Library, Galveston), but the lithographic copy was not published in his *St. Louis' Isle, or Texiana* until after his death (see opposite page 69).

12. Buchanan, *Life and Adventures of Audubon*, 343; Andrew Forest Muir, ed., *Texas in 1837: An Anonymous, Contemporary Narrative*, 196, note 3.

13. Buchanan, ed., *Life and Adventures of Audubon*, 343–344.

14. Audubon, *The Birds of America, from Drawings Made in the United States and Their Territories* (1870) 4:218, 5:303; Audubon to Thomas M. Brewer, n.p., October 29, 1837, in Herrick, *Audubon the Naturalist* 2:166. Audubon, *Ornithological Biography* 4:xvi–xviii, 58–67, 74–80, 81–87, 111–117, 118–129, 188–197, 203–211, 353–358, 600–607, lists many of the birds that he saw in Texas. See also Audubon, *Birds of America* (1840–1844) 6:139–147, 7:97, 119; Buchanan, *Life and Adventures of Audubon*, 339; and Holley, *Texas*, 100.

15. For example, see Pauline A. Pinckney, *Painting in Texas: The Nineteenth Century*, 43–48, especially plate 19; and Muir, ed., *Texas in 1837*, plate 3 between pages 108–109. For information on the painting, see Davidson, ed., *Original Water-Color Paintings* 2: plate 347; and Low, *Index and Guide*, 139. For fences, see Terry G. Jordan, *German Seed in Texas Soil: Immigrant Farmers in Nineteenth-Century Texas*, 97, 164.

16. Ford, *Audubon: A Biography*, 357–358; Geiser, "Audubon in Texas," 129; Herrick, *Audubon the Naturalist* 2:168. Havell had finished half of the engravings for volume 4 by the time Audubon left for America. Audubon to John Bachman, London, July 9, 1836, and April 6, 1837; and Audubon to Lucy Audubon, New York, July 3, 1837; all in Corning, ed., *Letters* 2:126, 160, 165–166. Also, Fries, *Double Elephant Folio*, 113, 400.

17. Geiser, "Audubon in Texas," 129; and Fries, *Double Elephant Folio*, 103–105. Geiser suggests that Audubon's Texas specimens were lost in shipment, but at least some of them arrived and were sent on to Brewer. See Audubon to Brewer, London, October 29, 1837, bMS Am 1482 (122); and Audubon to Edward Harris, London, November 18, 1837, pfMS Am 21 (39); both in Audubon Papers, Houghton Library.

18. Ford, *Audubon: A Biography*, 357–358; Low, *Index and Guide*, 13, 16. Low, 169, 175, also tells which of Audubon's crowded birds Havell distributed onto different plates. See, for example, prints 399 and 414 for the birds from original painting 327.

19. C. Jackson, *Bird Etchings*, 238; Fiero, "Audubon the Artist," 51, 53, 59.

20. Davidson, ed., *Original Water-Color Paintings* 1: plate 17; Audubon to Victor Audubon, Charleston, January 14, 1834, in Corning, ed., *Letters* 2:6; Reynolds, *Audubon and His Sons*, 45–47; Waldemar H. Fries, "Joseph Bartholomew Kidd and the Oil Paintings of Audubon's *Birds of America*," *Art Quarterly* 26 (Autumn 1967): 339–346. Ford, *Audubon: A Biography*, 437–445, lists all Kidd bird portraits copied after Audubon.

21. Davidson, ed., *Original Water-Color Paintings* 1: plate 219.

22. Williams, "Robert Havell, Junior," 252; C. Jackson, *Bird Etchings*, 32, 245, 254; Owens, *Audubon*, [22–23].

23. Herrick, *Audubon the Naturalist* 1:403–404, 2:199–201; Reynolds, *Audubon and His Sons*, 57; C. Jackson, *Bird Etchings*, 206, 214; and C. Jackson, *Bird Illustrators*, 33, 63; Winterfield, "American Ornithology," 271–274; and Sitwell, Buchanan, and Fisher, *Fine Bird Books*, 20–21.

24. Audubon to John Bachman, London, August 14, 1837, in Corning, ed., *Letters* 2:175–176, 179; Ford, *Audubon: A Biography*, 358; Low, *Index and Guide*, 16, 82, 186–191; C. Jackson, *Bird Engravings*, 32, 245, 254.

25. Fries, *Double Elephant Folio*, 114–117, 140, 151–171, 452; Ford, *Audubon: A Biography*, 364 (quoting from a manuscript in the Yale Audubon collection that cannot now be located), 366. Fries estimates (140) that there were between 175 and 200 complete sets issued. There were portions of perhaps as many as 120 sets in circulation, sent to subscribers who did not complete the subscription. The price of $874 to $1,000 in 1839 would have translated to approximately $12,708 to $14,540 in 1991, but, of course, the price of the double elephant folio has increased much more than that, with the University of Edinburgh copy selling for $4,070,000 in April 1992. McCusker, "How Much Is That in Real Money?" 327, 332.

CHAPTER THREE: THE ROYAL OCTAVO EDITION

1. Welch, "Audubon and His American Audience," 163.

2. Ford, *Audubon: A Biography*, 359–360; Victor H. Cahalane, ed., *The Imperial Collection of Audubon Animals: The Quadrupeds of North America*, x; Audubon to John Bachman, Boston, December 8, 1839, in Corning, ed., *Letters* 2:226; Audubon to S. G. Morton, New York, September 9, 1839, in Morton Papers, American Philosophical Society; Ford, *Audubon: A Biography*, 359–360.

3. Fries, *Double Elephant Folio*, 115; Durant and Harwood, *On the Road*, 536; C. Sellers, *Market Revolution*, 355–356.

4. *Albion* (New York), n.s. 1, October 5, 1839, p. 398, col. 3; H. E. Scudder, ed., *Recollections of Samuel Breck with Passages from His Note-Books (1771–1862)*, 260–261; Durant and Harwood, *On the Road*, 536.

5. Fries, *Double Elephant Folio*, 115–125; Audubon to John Bachman, London, August 14, 1837, in Corning, ed., *Letters* 2:176.

6. John Kirk Townsend's *Narrative of a Journey across the Rocky Mountains, to the Columbia River, and a Visit to the Sandwich Islands, for Chili, etc.* was reprinted in England as *Sporting Excursions in the Rocky Mountains, Including a Journey to the Columbia River, and a Visit to the Sandwich Islands, Chili, etc.* See also Ford, *Audubon: A Biography*, 355, 364; Victor Audubon, quoted in Fries, *Double Elephant Folio*, 307; Audubon to Edward Harris, Edinburgh, December 19, 1838, in Samuel N. Rhoads, "Auduboniana,"

Auk 20 (October 1903): 383; Herrick, *Audubon the Naturalist* 2:176. Townsend and Chevalier are quoted in Sotheby's *Library of H. Bradley Martin: Magnificent Color-Plate Ornithology*, lot 219. See also Nicholas B. Wainwright, *Philadelphia in the Romantic Age of Lithography*, 50; William H. Goetzmann, *Exploration and Empire: The Explorer and the Scientist in the Winning of the American West*, 185; and Coues, *Birds of the Colorado Valley*, 626.

7. Audubon to John Kirk Townsend, Edinburgh, January 1, 1839, fMS Am 21.5 (3), in Audubon Papers, Houghton Library; Townsend to Victor Audubon, Philadelphia, May 29, 1840, in box 5, folder 264, Audubon Papers, Beinecke Rare Book and Manuscript Library, Yale University, New Haven, Conn.; Audubon to S. G. Morton, New York, September 9, 1839, in Morton Papers, American Philosophical Society; and Sotheby's *Library of H. Bradley Martin: John James Audubon, Magnificent Books and Manuscripts*, lot 34.

8. See comments by Jacquelyn Sheehan in the Audubon files of the Stark Museum of Art, Orange, Texas; Fries, *Double Elephant Folio*, 307; Low, *Index and Guide*, 16–18.

9. Comments by Jacquelyn Sheehan in the Audubon files of the Stark Museum of Art. The only change made in the later octavo issue of *Harris's Hawk* (Bowen 5) is that it became a vertical composition; no landscape was added. The octavo version of the *Marsh Hawk* (Bowen 26) has been simplified, with the top young hawk eliminated, so that the plate contains only the adult and one young. The octavo version of the *Indigo Bird*, which is called *Indigo Bunting* (Bowen 170), contains the same four birds, but the composition does appear to be somewhat wider. At least, some of the wild sarsaparilla leaves and branches have been eliminated so that the overall effect is not so crowded. *Townsend's Surf Bird* (Bowen 322), which is the octavo version of *Townsend's Sandpiper*, has been turned into a horizontal plate and a landscape added.

10. The royal octavo edition is described in detail in Sotheby's *Library of H. Bradley Martin: Audubon, Magnificent Books and Manuscripts*, items 34 and 35; Wainwright, *Romantic Age of Lithography*, 50, 53; Herrick, *Audubon the Naturalist* 2:211, note 2; 212–214; Ford, *Audubon: A Biography*, 370, 372. The unbound set at the Beinecke Library, Yale University, which belonged to Miss C. Crowninshield of Boston, an original subscriber, begins with blue covers and changes to grey.

At $1 per number, Audubon's "small work" was expensive. The final price of $100 to $120, depending upon binding, would translate to $1,566 to $1,879 in 1991. McCusker, "How Much Is That in Real Money?" 332. In 1843 Audubon said that a "comfortable" home in St. Louis rented for $500 per year. He complained about hotels charging $9 to $12 per week. Eggs cost $.05 per dozen. Audubon to family, St. Louis, March 28, 1843, in John Francis McDermott, comp. and ed., *Audubon in the West*, 37; M. Audubon, ed., *Audubon and His Journals* 1:452.

11. Audubon to S. G. Morton, New York, September 9, 1839, in Morton Papers, American Philosophical Society; C. Jackson, *Bird Etchings*, 32; Welch, "Audubon and His American Audience," 74; *National Gazette and Literary Register* (Philadelphia), September 13, 1832; Field, "Audubon's Lithograph of the Clapper Rail," 147–153; Stefanie A. Munsing, *Made in America: Printmaking, 1760–1860: An Exhibition of Original Prints from the Collections of the Library Company of Philadelphia and the Historical Society of Pennsylvania, April–June, 1973*, 35.

12. Audubon to Edward Harris, Edinburgh, December 19, 1838, in Rhoads, "Auduboniana," 383; Audubon to S. G. Morton, New York, September 9, 1839, in Morton Papers, American Philosophical Society. Chevalier was listed along with Audubon on the title page of the first five volumes, but he was not a copublisher. See Ruthven Deane, "An Unpublished Letter of John James Audubon to His Fam-

ily," *Auk* 25 (April 1908): 167n; Wainwright, *Romantic Age of Lithography*, 49–50, 53–57; Whitman Bennett, *A Practical Guide to American Nineteenth Century Color Plate Books*, rev. ed., 5; Audubon, *Birds of America* (1844) 7:352. See also Ford, *Audubon: A Biography*, 369, 464. There are numerous references to Townsend being paid a commission. See Victor Audubon's account at the Bank of North America in Philadelphia, in "Audubon Journals and Ledgers," vol. 1, MCZ F118 (26), in Audubon Papers, Houghton Library, for example.

13. Wainwright, *Romantic Age of Lithography*, 33, 49–50, 54; McGrath, "American Colorplate Books," 37–43; Andrew F. Cosentino, *The Paintings of Charles Bird King (1785–1862)*, 60–61; Herman J. Viola, *Thomas L. McKenney, Architect of America's Early Indian Policy, 1816–1830*, 272–277; Dance, *Art of Natural History*, 111–146; Welch, "Audubon and His American Audience," 76. The Transco Energy Company collection in Houston, Texas, contains a Bowen watercolor. See *Contemplating the American Watercolor: Selections from the Transco Energy Company Collection*, plate 5.

14. Bennett, *Practical Guide*, 99; McGrath, "American Colorplate Books," 32–33; Peter H. Marzio, "The Democratic Art of Chromolithography in America: An Overview," in *Art and Commerce: American Prints of the Nineteenth Century*; and Janet Flint, "The American Painter-Lithographer," in *Art and Commerce*, 79, 92, 127. See also Alois Senefelder, *A Complete Course of Lithography*, introduction by A. Hyatt Mayor.

15. Recent immigrant Wilhelm C. A. Thielepape, for example, would have established a lithographic business in San Antonio, Texas, in 1855, if he could have mastered his used press and the techniques of lithography; but he failed, and no successful lithographic business operated in San Antonio until after the Civil War. *San Antonio Zeitung*, January 6 and October 10, 1855.

16. Harry T. Peters, *Currier and Ives: Printmakers to the American People*, 4–5; Peter H. Marzio, "American Lithographic Technology before the Civil War," in John D. Morse, ed., *Prints in and of America to 1850: Sixteenth Annual Winterthur Conference, 1970*, 216–217.

17. Bamber Gascoigne, *How to Identify Prints: A Complete Guide to Manual and Mechanical Processes From Woodcut to Ink Jet*. The Amon Carter Museum collection contains an Edward Beyer drawing, *High Bridge at Portage, B. & N.Y.C.R.R., Wyoming County, N.Y.* (accession number 139.83), which has been "squared," or marked in a grid pattern, for transfer to the stone.

18. John H. Hammond and Jill Austin, *The Camera Lucida in Art and Science*, 30–36; "Birds of America Day Book," 25, 47, in "Audubon Journals and Ledgers," vol. 5, MCZ F118 (26), in Audubon Papers, Houghton Library; Victor Audubon to John Woodhouse Audubon, New York, August 17, 1840, box 2, folder 51, in Audubon Papers, Beinecke Library.

19. Senefelder, *Complete Course of Lithography*, 170n, had earlier written, in regard to transfer lithographs, "It is hardly necessary to observe that only the outlines, and not every line in the minuter parts of the drawing, ought to be copied; for notwithstanding the greatest care in tracing them on the stone, it would be impossible in many cases to avoid confusion." Thirteen of the drawings have glue or wax on them; two have bits of tracing paper still embedded in the glue.

20. Ford, *Audubon: A Biography*, 369; Herrick, *Audubon the Naturalist* 2:214. Audubon received only 210 copies of number 3, but Bowen made up for the short run in the next printing. He sets out the terms of his agreement with Bowen on page 3. John was paid a total of $84 for doing the drawings for numbers 1–12. Audubon paid his agents 5 to 10 percent; see "Birds of America Day Book," 2–3, 7–8, 29, 47, in Audubon Papers, Houghton Library; Audubon to Victor Audubon, Charleston, May 4, 1840, in Corning, ed., *Letters* 2:267; Howard Corning, ed., *Journal of John James Audubon Made While Obtaining Sub-*

scriptions to His Birds of America, 16. See also Henry G. Bohn to Audubon, London, February 28, 1840, box 3, folder 90, in Audubon Papers, Beinecke Library. Audubon had written him about representing the book in England. The cost of $.57 per number does not include other expenses, such as additional salaries, office supplies, travel expenses, agents' commissions, and miscellaneous costs that are hard to pin down. The initials of other Bowen artists who drew the birds on stone are "A. V.," "J. C.," and "C. P." Some identities that have been suggested are John Collins, who did a series of views of Burlington, and Charles Parsons. The drawing signed "W" might have been the work of John Frampton Watson or Abraham Woodside. Conversation with Bernard Reilly, Chief Curator, Prints and Photographs Division, Library of Congress, February 19, 1991.

21. While comparable figures are difficult to obtain, and the price per copy varied considerably depending upon the lithograph's technical complexity and the edition, Schoolcraft's six-volume *Historical and Statistical Information Respecting the History, Condition and Prospects of the Indian Tribes of the United States*, which was published between 1851 and 1857, offers some idea of competing prices. Bowen produced twenty-one mostly hand-colored lithographs (page size was approximately 13 by $9\frac{1}{2}$ in.) for Schoolcraft in an edition of twelve hundred, similar to what he had done for Audubon a few years earlier, at a cost of $.075 each. James Ackerman of New York printed twelve hundred copies of thirty-five lithographs at a cost of $.045 each, while Peter S. Duval of Philadelphia printed twelve hundred copies of nineteen lithographs for $.062 each. Peter H. Marzio, *The Democratic Art: Chromolithography, 1840–1900, Pictures for a 19th-Century America*, 29–30.

By comparison, Ackerman produced 3,450 sets of sixty-eight lithographs (image size approximately 4 by 7 in.), uncolored and of lesser artistic merit (twelve tinted views, eleven geological illustrations, eighteen illustrations of reptiles, nineteen illustrations of flowers, six illustrations of fossils, and two large maps) for Randolph B. Marcy's *Exploration of the Red River of Louisiana, in the Year 1852*, in 1854 for $1.30 per set, or less than $.02 ($.019) per lithograph. He reprinted 5,000 sets later in the year for $1.25 per set, or $.018 per lithograph. See miscellaneous treasury account 115908, item 295, for Marcy's first edition, and misc. treas. acc. 117369, item 881, for the second edition. Three years later, Duval charged $1,176.06 to print seventeen quarto plates (approximately $5\frac{1}{2}$ by $8\frac{1}{2}$ in. image size) of views for William H. Emory's *Report on the United States and Mexican Boundary Survey*, in an edition of 11,530, or $.006 per print. Sarony and Company of New York drew on the stone and printed 6,400 copies each of twelve chromolithographs to illustrate the Emory *Report* for a total of $4,608, or $.06 each. Needless to say, these later prints were not done in the smaller editions that Audubon needed, nor were they hand-colored. See misc. treas. acc. 127113, item 1532, for Duval's payment, and misc. treas. acc. 127338, item 817, for Sarony and Company's payment, all in Records of the Accounting Officers of the Department of the Treasury, Office of the First Auditor, Record Group 217, National Archives, Washington, D.C.

22. The first copies of the first number were sent to Audubon in New York on December 3, 1839. Chevalier began making shipments to agents on December 13. "Birds of America Day Book," 1, Peters, *Currier and Ives*, 14–15.

23. The color proofs are collection 11.1/7 in the Stark Museum of Art.

24. Audubon to John Bachman, Boston, December 8, 1839, and January 2, 1840, in Corning, ed., *Letters* 2:227, 229.

25. John Bachman to Audubon, Charleston, January 13, 1840, in Herrick, *Audubon the Naturalist* 1:210.

26. Reviews of the octavo edition of *Birds of America*, in *Southern Cabinet* 1 (January 1840), no page number given; and 1 (February 1840): 122.

27. *Albion*, n.s. 2, January 25, 1840, p. 31, col. 3; Audubon to Lucy Audubon, Edinburgh, December 21, 1826; Audubon to Dr. Henry McMurtry, Edinburgh, February 21, 1831; both in Corning, ed., *Letters* 1:8, 129. Gopnik, "Audubon's Passion," 100. See also Robert Darnton, *The Great Cat Massacre and Other Episodes in French Cultural History*, 223–224. Herrick, *Audubon the Naturalist* 2:403, lists American reprints of only volumes 1 (Philadelphia) and 2 (Boston) of the *Ornithological Biography*.

28. Audubon to Lucy Audubon, Baltimore, March 1, 1840, in Corning, ed., *Letters* 2:239.

29. Herrick, *Audubon the Naturalist* 2:211, lists the agents from the last page of the paper covers in which the octavo edition was originally issued. Names of other agents, and sometimes the percentages paid them, are listed in the "Birds of America Day Book" and "Audubon Journals and Ledgers" in the Audubon Papers, Houghton Library.

30. Review of *Birds of America* in *Southern Cabinet* 1 (April 1840): 252; Scudder, ed., *Recollections of Samuel Breck*, 260. Entries for August 11, 12, and 18, 1840, for example, in Corning, ed., *Journal Made While Obtaining Subscriptions*, 3, 4, 8; Philadelphia *Mercury*, quoted in *Saturday Courier*, November 11, 1843, p. 2, col. 2; Parke Godwin, "Audubon," in *Homes of American Authors; Comprising Anecdotical [sic], Personal, and Descriptive Sketches*, 5. See also Welch, "Audubon and His American Audience," 101–140; Audubon to Victor Audubon, Baltimore, March 1 and 9, 1840, Richmond, April 12 and 15, 1840, and Charleston, May 4, 1840, in Corning, ed., *Letters* 2:239, 250, 254, 260, 262–264, 267.

31. The *Brother Jonathan* article is reprinted in Harold Edward Dickson, ed., *Observations on American Art: Selections from the Writings of John Neal (1793–1876)*, 66–68. See also Irving T. Richards, "Audubon, Joseph R. Mason, and John Neal," *American Literature* 6 (1934): 122–140; Herrick, *Audubon the Naturalist* 1:281, 2:432. Audubon claimed to have met Daniel Boone, but there seems to be no evidence that he did, although he corresponded with the frontiersman. Ford, *Audubon: A Biography*, 89.

32. Audubon to Victor Audubon, November 24, 1839, in Corning, ed., *Letters* 2:224; Audubon to S. G. Morton, New York, September 11, 1839, in Morton Papers, American Philosophical Society; *Albion*, n.s. 2, January 25, 1840, p. 31, col. 3; p. 32, col. 2; February 1, 1840, p. 40, col. 1.

33. Victor Audubon, note, in letter from Lucy Audubon to Audubon, n.p., February 27, 1840, box 1, folder 34, in Audubon Papers, Beinecke Library; Audubon to Victor, Baltimore, February 23, 1840, in Corning, ed., *Letters* 2:234–235; Herrick, *Audubon the Naturalist* 2:216; Parke Godwin, *Commemorative Addresses: George William Curtis, Edwin Booth, Louis Kossuth, John James Audubon, William Cullen Bryant*, 180. See *Albion*, n.s. 2, March 21, 1840, 99; April 25, 1840, 139; May 16, 1840, 143; May 27, 1840, 209; May 30, 1840, 179; and September 5, 1840, 291; Welch, "Audubon and His American Audience," 118–123.

34. Audubon to John Bachman, Boston, December 8, 1839, New York, January 2, 1840, and Baltimore, February 15, 1840, in Corning, ed., *Letters* 2:226, 228–229, 232; *Albion*, n.s. 2, January 25, 1840, p. 31, col. 3; p. 32, col. 2; February 1, 1840, p. 40, col. 1; February 8, 1840, p. 47, col. 3; "Birds of America Day Book," 2–3, 5, in the Audubon Papers, Houghton Library; Victor Audubon to Audubon, Charleston, November 25, 1839, and New York, December 14, 1839, box 2, folder 49, in Audubon Papers, Beinecke Library; Audubon to S. G. Morton, New York, September 11, 1839, in Morton Papers, American Philosophical Society; Audubon to Victor Audubon, Philadelphia, March 24, 1840, in Audubon Papers, American Philosophical Society; Ford, *Audubon: A Biography*, 373–374.

35. "Birds of America Day Book," 3, 6, 14, 20, 28, 37, 38–39, 89 in the Audubon Papers, Houghton Library; Victor Audubon to Audubon, [New Orleans,] n.d., fragment, box 12, folder 51, in Audubon Papers, Beinecke Library.

36. Audubon to Lucy Audubon, Baltimore, March 1, 1840; and Audubon to John Bachman, New York, March 9, 1840; both in Corning, ed., *Letters* 2:239, 250. Audubon to the family, Baltimore, February 21, 1840, in Herrick, *Audubon the Naturalist* 2:216–217; Robinson C. Watters, "Audubon and His Baltimore Patrons," *Maryland Historical Magazine* 34 (June 1939): 138–143.

37. Victor Audubon, notes, in letter from Lucy Audubon to Audubon, February 24, 1840; and Lucy Audubon to Audubon, n.p., March 1, 1840 (both in box 1, folder 34); Victor to John Bachman, New York, March 9, 1840 (box 2, folder 50); all in Audubon Papers, Beinecke Library. Also, note of W. J. (illegible) in Audubon to Victor, Philadelphia, March 24, 1840, in Audubon Papers, American Philosophical Society. Bowen delivered three hundred additional copies of number 1 on January 18, 1840, charging $27 per hundred to color the first two hundred and $28 to color the last hundred, and two hundred copies of number 2 on January 11, for which he charged $27 per hundred. See "Birds of America Day Book," 7–8, 20, 22; and Audubon to Maria Martin, Baltimore, February 29, 1840, bMS Am 1482 (151); both in the Audubon Papers, Houghton Library.

38. Audubon to Victor Audubon, Baltimore, March 9, 1840, Richmond, April 12 and 15, 1840, and Charleston, May 4, 1840, in Corning, ed., *Letters* 2:250, 254, 260, 262–264, 267.

39. Corning, ed., *Journal Made While Obtaining Subscriptions*, 3–27.

40. Ibid., 27–65.

41. Corning, ed., *Journal Made While Obtaining Subscriptions*, 4, 6, 7–8, 10–12, 14–18, 24, 47, 56–57; Ford, *Audubon: A Biography*, 373; Robert K. Shaw, "Elihu Burritt—Friend of Mankind," *Proceedings of the American Antiquarian Society*, 24–37. Also, Audubon to Edward Harris, Philadelphia, December 22, 1840, pfMS Am 21; and "Audubon Journals and Ledgers," vol. 1, MCZ F118 (215, 226–227); both in Audubon Papers, Houghton Library. Audubon received 896 copies of numbers 87 through 100, which probably means that he distributed a similar number of subscriptions throughout the publication. Thirty-one agents, distributing from 1 to 23 copies, received the remaining 154 copies.

42. "Notice of Audubon's *Birds of America*," *American Journal of Science and Arts* 42 (April 1842): 131; "Salvator" quote in Corning, ed., *Journal Made While Obtaining Subscriptions*, 99.

CHAPTER FOUR: A GREAT NATIONAL WORK

1. Victor Audubon, note, in letter from Lucy Audubon to Audubon, n.p., February 27, 1840, box 1, folder 34, in Audubon Papers, Beinecke Library.

2. Audubon to Dr. Henry McMurtry, Edinburgh, February 21, 1831, in Corning, ed., *Letters* 1:129; Audubon to S. G. Morton, New York, September 9, 1839, in Morton Papers, American Philosophical Society. Victor Audubon to Audubon, New York, March 2 and 6, 1840, box 2, folder 50; and Victor, note, in Lucy Audubon to Audubon, March 4, 1840, box 1, folder 34; all in Audubon Papers, Beinecke Library. Also, Audubon to Victor, Baltimore, March 7, 1840, in Corning, ed., *Letters* 2:250.

3. Audubon to Victor Audubon, Baltimore, March 7, 1840, in Corning, ed., *Letters* 2:246; Audubon to Victor, Philadelphia, March 24, 1840, New York, January 9, 1841, in Morton Papers, American Philosophical Society; Victor to Audubon, New York, March 2 and 6, 1840, box 2, folder 50, in Audubon Papers, Beinecke Library. See also Viola, *McKenney*, 274.

Notes to Pages 75-79

4. Audubon to family, February 21, 1841, in Herrick, *Audubon the Naturalist* 2:216, says that John and Maria have probably reached Charleston by then. Victor Audubon to John Woodhouse Audubon, New York, July 16, 1840, box 2, folder 50, in Audubon Papers, Beinecke Library. See chapter 3, note 20, for possible identities of "W."

5. Audubon to John Woodhouse Audubon, Baltimore, March 9, 1840, and Richmond, April 12, 1840, bMS Am 1482 (152, 153), in Audubon Papers, Houghton Library. Victor Audubon to John Bachman, New York, March 9, 1840, box 2, folder 50; Victor to John W. Audubon, New York, April 1 and 23, 1840, box 2, folder 50; all in Audubon Papers, Beinecke Library.

6. Victor Audubon, note, in letter from Eliza Audubon to John W. Audubon, New York, June 16, 1840, box 1, folder 6; Victor to Audubon, New York, July 16, 1840, box 2, folder 50; and Victor to John, New York, July 23, 1840, box 2, folder 51; all in Audubon Papers, Beinecke Library. Also, Audubon to Victor, New Bedford, July 30, 1840, in Corning, ed., *Letters* 2:278.

7. Eliza Audubon to John W. Audubon, New York, August 2, 1840, box 1, folder 6; John to Victor Audubon, August 10, 1840, box 1, folder 17; Victor to John, New York, August 17, 1840, box 2, folder 51; and John to Victor, Charleston, September 2, 1840, box 1, folder 17; all in Audubon Papers, Beinecke Library. Audubon gives September 15 as the date of Maria's death in his *Journal Made While Obtaining Subscriptions*, p. 25; Herrick, *Audubon the Naturalist*, 2:218, gives September 23 as the date.

8. Audubon to S. G. Morton, New York, October 12, 1840, in Morton Papers, American Philosophical Society; Corning, ed., *Journal Made While Obtaining Subscriptions*, 25, 75–76.

9. Parke Godwin, "John James Audubon," *United States Magazine and Democratic Review* 10 (May 1842): 449; "Birds of America Day Book," 120, 159, and 216, in the Audubon Papers, Houghton Library.

10. "Birds of America Day Book," 88.

11. Ibid., 159.

12. Victor Audubon to Audubon, New Orleans, November 20 and 28 and December 12, 1840, undated letter fragment, and January 9, 1840 [1841], box 12, folder; 51 Havana, January 3, 1841, Cangrejo (near Matanzas), March 12, 1841, box 2, folder 52; all in Audubon Papers, Beinecke Library. Audubon to children, New York, January 27, 1841, 11.1/101, in Audubon Papers, Stark Museum of Art; "Birds of America Day Book," 42, 45, 50, 55, 57, 62, 64, 68, 70, 74, 78, 81, 85, 89, in the Audubon Papers, Houghton Library.

13. Victor Audubon to Audubon, San Pedro, Cuba, February 23, 1841, box 2, folder 52, in Audubon Papers, Beinecke Library; Herrick, *Audubon the Naturalist* 2:218. Audubon wrote Edward Harris, Philadelphia, December 22, 1840, pfMS Am 21, in Audubon Papers, Houghton Library, that he would be at home "the whole of the winter."

14. "Birds of America Day Book," 91, 94, Audubon Papers, Houghton Library; Audubon to Victor Audubon, New York, January 30, 1841, January 9, 1841, and February 11, 1841, in Morton Papers, American Philosophical Society; Audubon to Little and Brown, New York, April 29, 1841, quoted in Herrick, *Audubon the Naturalist* 2:230.

15. Audubon probably did not distribute all of the plates with block lettering or the incorrect titles first; he probably began using the corrected plates as soon as he received them. Bowen had delivered the second printing of numbers 1 through 5 by March 26, 1840 (see Table 1). Audubon might have used the new printings to make up his sets at that point, putting the remainder of the earlier printings back for use later, if needed. A comparison of two sets in the Beinecke Rare Book and Manuscript Library at

Yale University suggests that this might have happened. One set, still in original wrappers, belonged to Miss C. Crowninshield of Boston (call number Sy13/A12/+840A/3), the other, which is bound, to Edmund Penfold of New York City (1975/2340/1). Since Audubon published the *Birds* in numerical order and listed his subscribers on the covers, one can ascertain roughly when the persons listed subscribed. For example, Crowninshield's name first appears on the cover of number 11 in a column labeled "Subscribers since No. 6," which probably means that she subscribed sometime after February 7, 1840, when the covers for number 6 were printed. Edmund Penfold's name appears in the subscriber list with number 6, which means that his subscription predated February 7. But Crowninshield's copy has both the block lettering in the first number and the incorrect titles in number 5 that are characteristic of the earlier printings, while Penfold's copy has neither. It is likely that Audubon received the corrected copies of numbers 1 through 5 in time to supply them to Penfold, but that the number of subscribers had increased so rapidly that he ran out of them by the time Crowninshield subscribed, so he had to give her the earlier versions. See "Birds of America Day Book," 38, 71, Audubon Papers, Houghton Library.

Three copies of the first edition of *Birds* at the Beinecke Library, the Crowninshield, Penfold, and Amos Lawrence (Sy13/A12/840Ab/1) copies, show the birds in plate 16 close together; a fourth copy (1974/2885/1) shows them three-quarters of an inch apart.

16. Audubon to Victor Audubon, New York, January 30, 1841, in Morton Papers, American Philosophical Society; Victor, note, May 21, 1844, in letter from Lucy Audubon to Audubon, n.p., n.d., box 1, folder 38, in Audubon Papers, Beinecke Library. W. Thomas Taylor of Austin, Texas, acquired the incomplete set from the U.S. Department of Agriculture.

17. "Birds of America Day Book," Audubon Papers, Houghton Library. When the project was fully underway, Bowen was producing so many lithographs that he requested that Audubon pay a portion of the cost in advance, so Bowen would not have to bear all the financial burden. See 57, 59, 66, 68, and 70 of the "Day Book" for examples. Burgess quote in Ford, *Audubon: A Biography*, 395.

18. Audubon to John W. Audubon, New York, January 25, 1841, in Albert E. Lownes, "Ten Audubon Letters," *Auk* 52 (April 1935): 162.

19. Ford, *Audubon: A Biography*, 384; Viola, *McKenney*, 273–275.

20. Audubon to Victor Audubon, in Lucy Audubon to Victor, New York, February 24, 1841, box 1, folder 35; and George Endicott to Audubon, New York, April 26 and July 13, 1841, box 3, folder 117; all in Audubon Papers, Beinecke Library. Ford, *Audubon: A Biography*, 384, 493, says that Childs charged Audubon $1,900 per hundred numbers (of five prints each). Surely that is incorrect, since that would result in a charge of $3.80 per print, when Audubon was selling them for $1 per number.

It is also possible that, because he was printing fifteen hundred copies of each image, Endicott made more than one stone of each bird and that the proof of the *Common Mocking Bird* in the Hill Memorial Library collection is a version for which I have not yet seen a published image. This is probably not the case, since I have seen no variations in the published work, other than the *Black-shouldered Elanus* already noted, that are so different.

21. Wainwright, *Romantic Age of Lithography*, 57.

22. Glyndon G. Van Deusen, *The Jacksonian Era, 1828–1848*, 155–164; Edward Pessen, *Jacksonian America: Society, Personality, and Politics*, 49, 96; Robert V. Remini, *Andrew Jackson and the Course of American Democracy, 1833–1845* 3:427–433; Audubon to Victor Audubon, Baltimore, March 7, 1840, in Corning, ed., *Letters* 2:247. Victor to Audubon, New York, March 6, 1840, box 2, folder 50; and

undated letter fragment, box 12, folder 51; both in Audubon Papers, Beinecke Library. Also, Audubon to Victor, January 9, 1841, in Morton Papers, American Philosophical Society.

23. John W. Audubon to Audubon, Philadelphia, June 12, 1841, and Victor Audubon, note, box 1, folder 17; Lucy Audubon to Audubon, New York, December 20, 1842, box 1, folder 36; Victor, note, New York, July 14, 1843, in letter from Lucy Audubon to Audubon, Minnie's Land, July 2, 1843, box 1, folder 37; and Victor, note, in letter from Lucy Audubon to Audubon, n.p., July 28, 1844, box 1, folder 40; all in Audubon Papers, Beinecke Library. Audubon to Victor, January 9 and February 11, 1841, February 12, 1842, July 12, 1842, all in Morton Papers, American Philosophical Society. S. G. Morton to John Bachman, Philadelphia, January 28, 1843, bMS Am 1482 (364), in Audubon Papers, Houghton Library.

24. Victor Audubon, note, in letter from Lucy Audubon to Audubon, New York, July 21, 1842, box 1, folder 36, in Audubon Papers, Beinecke Library; Audubon to Spencer Baird, New York, July 29, 1841, in Herrick, *Audubon the Naturalist* 2:227; Audubon to John Bachman, Minnie's Land, November 12, 1843, bMS Am 1482 (160), in Audubon Papers, Houghton Library; Ford, *Audubon: A Biography*, 395-398.

25. Herrick, *Audubon the Naturalist* 2:214; Coues, *Birds of the Colorado Valley*, 630; Bannon and Clark, *Handbook of Audubon Prints*, 41-43, 91-108 (listing all of the 435 prints in the original *Birds of America*, including the date and place of their painting, none of them having been painted in Texas, or from Texas specimens); C. S. Francis and Co., New York, advertisement for the "Works of John James Audubon," in Hill Memorial Library; Audubon, *Birds of America* (1844) 7:352.

26. *The Eagle of Washington* (1837, oil on canvas, Mr. and Mrs. Dillon Ripley and Mrs. Gerald M. Livingston), illustrated in David Carew Huntington, *Art and the Excited Spirit: America in the Romantic Period*, plate 33.

27. Low, *Index and Guide*, 185-191.

28. Huth, *Nature and the American*, 25, says that Wilson was far better known than Audubon until after publication of the octavo edition. See also Welch, "Audubon and His American Audience," 141, 147-151, 152-157, 161-165; *New Orleans Directory for 1842*, 52; John Rewald, "Degas and His Family in New Orleans," in *Edgar Degas, His Family and Friends in New Orleans*, John Rewald et al., eds., 11; *Saturday Courier*, May 20, 1843, p. 2, col. 5; *Albion*, September 5, 1840, p. 291, col. 3, and July 15, 1843, p. 353, col. 3. For a discussion of America's need for national works, see Richard Rudisill, *Mirror Image: The Influence of Daguerroetype on American Society*, 5-7.

29. Welch, "Audubon and His American Audience," 195-200.

30. Audubon sold the double elephant folio for approximately $10 per number, or $1,000 per set. He wound up with 161 subscribers and sold fifteen additional sets for approximately $1,000 to $1,100 each, depending upon the binding, after returning to the United States in 1839. His total income would have been somewhere in the neighborhood of $150,000 to $155,000, with expenses of $115,640, for a potential profit of approximately $35,000 to $40,000. Fries, *Double Elephant Folio*, 114-117,, 151-171, 198.

Potential profit for the octavo edition is derived from figures in the "Birds of America Day Book" and "Audubon Journals and Ledgers," both in the Audubon Papers, Houghton Library. By the end of the publication, Audubon was distributing 896 copies of each number, and thirty-one agents got 154 copies.

31. Yarrell to Audubon, March 10, 1841, quoted in Herrick, *Audubon the Naturalist* 2:223-224; Peter H. Marzio, *Mr. Audubon and Mr. Bien: An Early Phase in the History of American Chromolithography*, [4];

Saturday Courier, August 19, 1843, p. 3, col. 3; November 25, 1843, p. 2, col. 6; *Albion*, May 27, 1840, 209.

32. Welch, "Audubon and His American Audience," 236–237, 240–241, 244, 246–247.

CHAPTER FIVE: SUBSEQUENT EDITIONS OF *THE BIRDS OF AMERICA*

1. Victor Audubon, note, March 20, 1843, in letter from Lucy Audubon to Audubon, March 19, 1843, box 1, folder 36; Victor, note, April 3, 1843, in Lucy Audubon to Audubon, April 2, 1843, box 1, folder 36; Victor, note, April 10, 1843, in Lucy Audubon to Audubon, April 9, 1843, box 1, folder 36; and Victor, note, June 2, 1843, in Lucy Audubon to Audubon, Minnie's Land, May 29, 1843, box 1, folder 37; all in Audubon Papers, Beinecke Library. Ford, comp. and ed., *Quadrupeds*, 59.

2. Quote from *Auk*, 1917, given in Ford, *Audubon: A Biography*, 401; see also 402–412. Audubon's journal of the trip can be found in M. Audubon, ed., *Audubon and His Journals* 1:452–532, 2:1–195. His letters home are published in McDermott, comp. and ed., *Audubon in the West*, and Edward Harris's journal is in McDermott, ed., *Up the Missouri with Audubon: The Journal of Edward Harris*.

3. M. Audubon, ed., *Audubon and His Journals* 2:27, 96, 108.

4. Audubon to John Bachman, Minnie's Land, December 10, 1843, bMS Am 1482 (161); Audubon to Bachman, Minnie's Land, February 1, 1846, bMS Am 1482 (165), and Victor Audubon, note, New York, March 1, 1846, in letter from Audubon to Bachman, Minnie's Land, March 1, 1846, bMS Am 1482 (167); all in Audubon Papers, Houghton Library. Also, Victor, note, New York, June 21, 1844, in Lucy Audubon to Audubon, n.d., box 1, folder 38, in Audubon Papers, Beinecke Library; *Northern Standard* (Clarksville, Texas), March 11, 1846, p. 1, col. 4, quoting a December 25, 1845, story from the *Galveston News*; Ford, *Audubon: A Biography*, 416, 418, 421; Herrick, *Audubon the Naturalist* 2:261–262.

5. Bannon and Clark, *Handbook of Audubon Prints*, 70–72.

6. Bannon and Clark, *Handbook of Audubon Prints*, 71–76; Herrick, *Audubon the Naturalist* 2:293. In 1854 a supplement of ninety-three pages of text and six hand-colored lithographs apparently was issued to all subscribers of the imperial folio *Quadrupeds*. It is sometimes bound with volume 3 of the text. The six plates were also included in the octavo edition, the *Mountain Brook Mink* (Octavo 124) and the last five plates in the book (Octavo 151–155). Bennett, *American Nineteenth Century Color Plate Books*, 5.

7. Fries, *Double Elephant Folio*, 123, citing MS 311, in collections of letters written by and to John James Audubon, given to the Museum of Comparative Zoology, Harvard University, by Colonel John E. Thayer; deposited in the Houghton Library, Harvard University. The partial set of first edition text with second edition plates consisting of volumes 1, 2, 3, and 4 was obtained as a duplicate from the U. S. Department of Agriculture and was in possession of W. Thomas Taylor of Austin, Texas.

8. John Cassin to Victor Audubon, Philadelphia, June 9, 1851, box 3, folder 98; and Victor to John W. Audubon, Charleston, March 2, 1852, box 2, folder 58; both in Audubon Papers, Beinecke Library. Also, Victor to Cassin, New York, June 4, 1852, with attachment that might be a copy of Cassin's terms, pfMS Am 21 (78), in Audubon Papers, Houghton Library.

9. Victor Audubon to John Cassin, New York, June 4, 1852, with attachment that might be a copy of Cassin's terms, pfMS Am 21 (78), in Audubon Papers, Houghton Library.

10. Robert McCracken Peck, Introduction to the facsimile reprint of John Cassin, *Illustrations of the*

Birds of California, Texas, Oregon, British and Russian America, I-4, I-34; Robin W. Doughty, *The Mockingbird*, 64–65; George Ord to Charles Waterton, Philadelphia, June 22, 1845, in Audubon Papers, American Philosophical Society; Victor Audubon to John Bachman, New York, September 4, 1852, bMS Am 1482 (305), in Audubon Papers, Houghton Library.

11. Marzio, *Democratic Art*, 29–30; Charles Webber's *The Hunter-Naturalist: Romance of Sporting; or, Wild Scenes and Wild Hunters* appeared in 1851; *The Hunter-Naturalist: Wild Scenes and Song-Birds* in 1852. See Warder H. Cadbury, "Alfred Jacob Miller's Chromolithographs," in Ron Tyler, ed., *Alfred Jacob Miller: Artist on the Oregon Trail*, 447–448; Cassin, quoted in Peck, Introduction, I-17. See also Chapter 3, note 15.

12. James Gilreath, "American Book Distribution," in David D. Hall and John B. Hench, eds., *Needs and Opportunities in the History of the Book: America 1639–1876*, 137–138; Peck, Introduction, I-14, I-39; Victor Audubon to John Bachman, New York, May 21, 1852, bMS Am 1482 (297), in Audubon Papers, Houghton Library.

13. Victor Audubon to John Bachman, New York, June 14, 1852, bMS Am 1482 (298); and Victor to Bachman, New York, June 25, 1852, bMS Am 1482 (299); both in Audubon Papers, Houghton Library.

14. Peck, Introduction, I-18, I-20.

15. Ibid., I-20–21.

16. Ibid., I-3–4, I-21–22.

17. Ibid., I-21; John Cassin to Victor Audubon, Philadelphia, June 4, 1855, box 3, folder 98, in Audubon Papers, Beinecke Library.

18. Audubon to Victor Audubon, New York, September 23, 1833, in Corning, ed., *Letters* 1:259; Victor to Audubon, n.p., August 7, 1843; Victor to John Bachman, in Fries, *Double Elephant Folio*, 123, re. having sold some copies of the "little work" that will have to be printed. The practice of keeping the extra letterpress led to at least one partial set of books (volumes 1–4), in the possession of W. Thomas Taylor of Austin, Texas, with first edition text and second edition plates. Audubon's injunction to Victor to keep the stones related to his desire to begin the project in secret; keeping the stones in his possession would have helped guard the secret until the publication was announced and sold. Bills from the printer, E. G. Dorsey, noted in the "Birds of America Day Book," 6, 115, Audubon Papers, Houghton Library, show the number of prospectuses, covers, and texts that Audubon had printed.

19. For these conclusions I compared the first edition in the collection of the Texas Memorial Museum with the second edition in the collection of the Harry Ransom Humanities Research Center, both on the campus of the University of Texas at Austin.

20. Journal entry for June 25, 1856, in Bradford Torrey and Francis H. Allen, eds., *The Journals of Henry D. Thoreau* 8:287; Herrick, *Audubon the Naturalist* 2:294. The publishing history of the octavo *Quadrupeds* is even more confusing than that of the small *Birds*. Ford, *Audubon's Animals*, 217, says that it was reprinted in 1854, 1856, and, perhaps, 1860. Bannon and Clark, *Handbook of Audubon Prints*, 76, agree on the number of printings but list 1854, 1856–1860, and 1870 as the dates. The *National Union Catalogue* lists five printings: 1849–1854, 1854–1855, 1856, 1856–1860, and 1870. It, too, has been reprinted at least twice in this century: in the Volair limited edition (1979) and on CD-ROM.

21. Peck, Introduction, I-28, I-30.

22. Ibid., I-28, I-32 (Cassin died in 1869).

23. I am indebted to David M. Lank of Montreal for sharing his notes on the comparison of the plates

in Audubon's first edition (1840–1844) with those in the third edition (1859); much of this information comes from his notes. I also had the opportunity to examine third edition copies at Rice University in Houston (a mixed set including 1859 and 1860 imprints) and Southern Methodist University in Dallas. Victor suffered his injury in a fall, although Herrick, *Audubon the Naturalist* 2:295, was not sure whether it occurred at home or in a railroad accident.

24. Audubon, *Birds of America* (1840) 1:119, 141; Carl Shaefer Dentzel in *The Drawings of John Woodhouse Audubon, Illustrating His Adventures Through Mexico and California, 1849–1850*, [8], plate 10. The changes between the second and third editions of the octavo *Birds of America* have been confirmed by examining copies in the Harry Ransom Humanities Research Center (second edition) and the Science Library at Southern Methodist University (third edition); Bannon and Clark, *Handbook of Audubon Prints*, 50; Herrick, *Audubon the Naturalist* 2:295. John Woodhouse Audubon also began production of a series of prints based on his Mexico and California drawings, but he did not complete the work. These prints can be found in *Illustrated Notes of an Expedition Through Mexico and California*. For a discussion of how changes were made to lithographic stones, see Michael Twyman, *Lithography, 1800–1850: The Techniques of Drawing on Stone in England and France and Their Application in Works of Topography*, 136–137.

25. A watercolor in the collection of the Museum of Comparative Zoology at Harvard University may be the original study for the octavo illustration of the *Wild Turkey Hen*. It seems a bit large, $9\frac{3}{4}$ by $13\frac{3}{4}$ inches, and it shows eight chicks rather than the six in all versions of the octavo print. Ford, *Audubon: A Biography*, 164.

26. Herrick, *Audubon the Naturalist* 2:268, 295; Ford, *Audubon: A Biography*, 416; Fries, *Double Elephant Folio*, 391–392. Charles Winterfield, "A Talk About Birds," *American Review: A Whig Journal* 2 (September 1845): 287, describes the fire and its impact.

27. Herrick, *Audubon the Naturalist* 2:296–297; Marzio, *Democratic Art*, 51, 53, 55; agreement between John W. Audubon and Julius Bien, November 7, 1859, with George R. Lockwood as witness, 11.2/2, in Audubon Papers, Stark Museum of Art.

28. Herrick, *Audubon the Naturalist* 2:389; Fries, *Double Elephant Folio*, 355.

29. Agreement between Audubon and Bien, November 7, 1859, in Stark Museum of Art, Prospectus for the Bien edition, issued through Messrs. Trübner & Company, London, 1859, quoted in Herrick, *Audubon the Naturalist* 2:389; Marzio, *Democratic Art*, 56–57.

30. Marzio, *Democratic Art*, 57.

31. Ibid., 73.

32. Although there are illustrated copies of the 1861 edition bound in seven volumes at Cornell University and the Brooklyn Museum (see RLIN citations NYCX85-B26364 and NYBA90-B7271), most of them are bound in five volumes and are not illustrated, suggesting that they were intended as the text to accompany the Bien edition. See OCLC entry number 680167.

33. William Brotherhead, *Forty Years Among the Old Booksellers of Philadelphia, with Bibliographical Remarks*, 15–17; Marzio, *Democratic Art*, 55; Herrick, *Audubon the Naturalist* 2:296; John W. Audubon to J. T. Johnston, New York, November 5, 1861, 11.2/105, in Audubon Papers, Stark Museum of Art.

34. Marzio, *Democratic Art*, 55; Herrick, *Audubon the Naturalist* 2:295–297; Fries, *Double Elephant Folio*, 125, 392, 394; Audubon to John Bachman, Minnie's Land, March 12, 1846, bMS Am 1482 (168), in Audubon Papers, Houghton Library.

35. Fries, *Double Elephant Folio*, 306–309, 390–398; Herrick, *Audubon the Naturalist* 2:407; John W. Audubon to J. T. Johnston, New York, November 5, 1861, 11.2/105, in Audubon Papers, Stark Mu-

seum of Art. John offered Johnston *The Birds of America*, *The Viviparous Quadrupeds of North America*, and John Gould's *Birds of Europe* (5 vols.; London, 1832–1837), *Birds from the Himalaya Mountains* (London, 1831–1832), and his *Monograph of the Ramphastidae, or Family of Toucans* (London, 1833–1835), all for $1,200. Johnston paid $1,000 for the group. Joel J. Orosz, *Curators and Culture: The Museum Movement in America, 1740–1870*, 215. See also *Audubon's Birds of America: Life-Size Drawings from the Original Stones.* . . . Copies of this catalog exist in at least two collections, the Beinecke Rare Book and Manuscript Library at Yale and the New-York Historical Society. The historical society copy has "1883" written in a contemporary hand at the bottom of the cover. See RLIN citations CTYX83-B29 and NYHR88-B4330.

36. Richard B. Lockwood's statement is handwritten on the first blank page of a sales sample copy that combines sample plates and text of the *Birds of America* and the *Quadrupeds of North America* in one volume, in the collections of the American Antiquarian Society, Worcester, Massachusetts. A page from *Publishers Weekly*, no. 1803, August 18, 1906, 368, laid in the copy, contains an obituary of George R. Lockwood which states, "The lithographic stones from which these [octavo] editions were printed were destroyed in a fire that consumed the building in Philadelphia where they were stored." Christie's, *Valuable Travel and Natural History Books*, lot 219, cites the 1889 edition.

37. B. H. Warren, *Report on the Birds of Pennsylvania, with Special Reference to the Food-Habits, Based on over Four Thousand Stomach Examinations*, vii.

CHAPTER SIX: AUDUBON IN AMERICAN ART

1. See, for example, Fries, *Double Elephant Folio*, 126–133; James Thomas Flexner, *That Wilder Image: The Painting of America's Native School from Thomas Cole to Winslow Homer*, 86, 277; Theodore E. Stebbins, Carol Troyen, and Trevor J. Fairbrother, *A New World: Masterpieces of American Painting, 1760–1910*, 340; John Wilmerding, *American Art*, 124–125, and *Audubon, Homer, Whistler and 19th Century America*, 9–10; Joshua Taylor, *America As Art*; and quote in *Cincinnati Daily Gazette*, November 15, 1833, p. 2, col. 3.

2. Ford, *Audubon: A Biography*, 260, 287–289; Audubon to Lucy Audubon, Charleston, November 7, 1831, in Corning, ed., *Letters* 1:148; "Report . . . to Examine Splendid Work of Mr. Audubon," *Silliman's Journal* 16 (1829): 353–354.

3. Dunlap, *Rise and Progress* 3:204–205; Henry T. Tuckerman, "Birds and Audubon," *Methodist Quarterly Review* 34 (July 1852): 422, 429; Henry T. Tuckerman, *Book of the Artists: American Artist Life Comprising Biographical and Critical Sketches of American Artists: Preceded by an Historical Account of the Rise and Progress of Art in America*, 494–495.

4. Earl A. Powell, *Thomas Cole*, 12–26, 71, 73.

5. Barbara Novak, *American Painting of the Nineteenth Century: Realism, Idealism, and the American Experience*, 136; James Thomas Flexner in *That Wilder Image*, 86–87, *History of American Painting, 1760–1835: The Light of Distant Skies* 2:227, and *America's Old Masters: First Artists of the New World*, 229 (mention only); Milton W. Brown, *American Art: Painting, Sculpture, Architecture, Decorative Arts, Photography*, 224–225; and Oliver W. Larkin, *Art and Life in America*, 110. John Wilmerding in *American Art*, 124;

Wilmerding, ed., *The Genius of American Painting*, 126–127; and Theodore E. Stebbins, *American Master Drawings and Watercolors: A History of Works on Paper from Colonial Times to the Present*, 74–81.

6. Audubon, quoted in Wilmerding, *American Art*, 125. Wilmerding quotes from Ford, ed., *Audubon, by Himself*, 223–224. See also M. Audubon, ed., *Audubon and His Journals* 2:199, 204, 237, 262, 206; Audubon, *Birds of America* (1840) 1:92; Welker, *Birds and Men*, 105.

7. Fiero, "Audubon the Artist," 53; Audubon, Introduction to *Ornithological Biography* 1:xii.

8. Charles Rosen and Henri Zerner, *Romanticism and Realism: The Mythology of Nineteenth Century Art*, 24–28, 74–84.

9. Emerson, quoted in Huth, *Nature and the American*, 88; Welker, *Birds and Men*, 91, 96–97.

10. Welker, *Birds and Men*, 85–86, 110. A recent publication on the meaning and significance of genre painting is Elizabeth Johns, *American Genre Painting: The Politics of Everyday Life*.

11. See Welker, *Birds and Men*, 131–132; Owens, *Audubon*, [22]; Audubon, *Birds of America* (1844) 4:110–111 (blue jay), 413 (cormorant); Ford, ed., *1826 Journal*, 250 (great-footed hawk).

12. C. Jackson, *Bird Etchings*, 241.

13. Audubon, *Ornithological Biography* 1:60–61; Huntington, *Art and the Excited Spirit*, 3, 21–22; Audubon, *Birds of America* (1840) 1:55 (bald eagle); Ford, *Audubon: A Biography*, 301.

14. William H. Truettner, *The Natural Man Observed: A Study of Catlin's Indian Gallery*, 20–21, 105, 184; Dippie, *Catlin and His Contemporaries*, 30.

15. Army R. W. Meyers, "Imposing Order on the Wilderness: Natural History Illustration and Landscape Portrayal," in Edward J. Nygren and Bruce Robertson, *Views and Visions: American Landscape before 1830*, 120–121; C. C. Sellers, *Mr. Peale's Museum*, 18–19; Audubon to Lucy Audubon, Edinburgh, March 24, 1827, in Corning, ed., *Letters* 1:19. Audubon hoped that John would be able to provide new drawings for Havell's press. Audubon to Lucy Audubon, London, August 8, 1828, in Corning, ed., *Letters* 1:67.

16. Ford, *Audubon: A Biography*, 262; Donald C. Peattie, ed., *Audubon's America: The Narratives and Experiences of John James Audubon*, 184. Bannon and Clark, *Handbook of Audubon Prints*, 92–108, list the assistants, including Victor, who helped Audubon and the images to which they contributed.

17. George Catlin, *Letters and Notes on the Manners, Customs, and Conditions of North American Indians* 1:3; M. Audubon, ed., *Audubon and His Journals* 1:407; Lee Clark Mitchell, *Witness to a Vanishing America: The Nineteenth-Century Response*, 35–36; Annette Kolodny, *The Lay of the Land: Metaphor As Experience and History in American Life and Letters*, 74–88, especially 74–77; "The Ohio," in Audubon, *Ornithological Biography* 1:31–32. Audubon also published the essay on the Ohio in *Albion*, May 27, 1840, 209–210.

18. Welch, "Audubon and His American Audience," 211–212, 222. Parkman's copy of the *Ornithological Biography* is in the collection of the Boston Athenaeum. See 2:154–159, especially 154. See also George C. Shattuck to Audubon, Boston, July 25, 1844, box 5, folder 247, in Audubon Papers, Beinecke Library; *Saturday Courier*, November 11, 1843, p. 2, col. 2; *Albion*, March 9, 1844, 119. For William Cullen Bryant, "A Forest Hymn," see Taylor, *America As Art*, 104.

19. Herrick, *Audubon the Naturalist* 1:360; Winterfield, "American Ornithology," 274. Winterfield did not go into the "wild natural scenes" in his next essay, perhaps because a fire destroyed some of Audubon's copperplates in the meantime, and Winterfield concluded his article with a comment on the tragedy (he erroneously thought they had all been destroyed). See Winterfield, "A Talk about Birds," 279–287. See also John W. McCoubrey, *American Tradition in Painting*, 23–31, which discusses

the importance of realism and landscape in American art, including Audubon's role. See also page 25.

20. The Philadelphia *Mercury*, quoted in *Saturday Courier*, November 11, 1843, p. 2, col. 2; R. B. Rhett to Audubon, n.p., February 4, 1841, box 5, folder 234, in Audubon Papers, Beinecke Library; Welch, "Audubon and His American Audience," 222, 236–243; Dawn Glanz, *How the West Was Drawn: American Art and the Settling of the Frontier*, 8.

21. William F. Stapp, "Daguerreotypes onto Stone: The Life and Work of Francis D'Avignon," in Wendy Wick Reaves, ed., *American Portrait Prints: Proceedings of the Tenth Annual American Print Conference*, 203–204, 213; Rudisill, *Mirror Image*, 19, 173.

22. *Saturday Courier*, August 19, 1843, p. 3, col. 3; November 11, 1843, p. 2, col. 2. Also, Gordon Hendricks, *The Life and Work of Winslow Homer*, 236; John Wilmerding, "Winslow Homer's *Right and Left*," *Studies in the History of Art* 9 (1980): 79–80; Ella M. Foshey, *Reflections of Nature: Flowers in American Art*, 149.

23. Sitwell, Buchanan, and Fisher, *Fine Bird Books*, 15; Theodore E. Stebbins, "Luminism in Context: A New View," 211–234, points out European antecedents for several styles of American art.

Bibliography

PRIMARY SOURCES
MANUSCRIPTS

Audubon, John James. Papers. American Philosophical Society, Philadelphia.
——. Papers. Beinecke Rare Book and Manuscript Library, Yale University, New Haven, Conn.
——. Papers. Houghton Library, Harvard University, Cambridge, Mass.
——. Papers. Stark Museum of Art, Orange, Texas.
Morton, S. G. Papers. American Philosophical Society, Philadelphia.
Records of the Accounting Officers of the Department of the Treasury, Office of the First Auditor. Record Group 217. National Archives, Washington, D.C.

PUBLISHED WORKS

Albion (New York). 1840–1844.
Annual II: Studies on Thomas Cole, an American Romanticist. Baltimore: Baltimore Museum of Art, 1967.
Audubon, John James. *The Art of Audubon: The Complete Birds and Mammals: John James Audubon*. Introduction by Roger Tory Peterson. New York: Times Books, 1979.
——. *Audubon's Animals: The Quadrupeds of North America*. Comp. and ed. by Alice Ford. New York: The Studio Publications, Inc., in association with Thomas Y. Crowell Company, 1951.
——. *Audubon's Birds of America*. Text by George Dock, Jr. New York: Arrowood Press, 1987.
——. *Audubon's Birds of America: Life-Size Drawings from the Original Stones....* Boston: Estes and Lauriat, 188[3].
——. *Audubon's Birds of North America*. Introduction by Sheila Buff. Stamford, Conn.: Longmeadow Press, 1990.
——. *Audubon's Quadrupeds of North America*. Foreword by William Kammer. Secaucus, N.J.: The Wellfleet Press, 1989.
——. *The Birds of America, from Drawings Made in the United States and Their Territories*. 7 vols. Vols.

191

1–5, New York: Published by J. J. Audubon and, Philadelphia: J. B. Chevalier, 1840–1843; Vols. 6–7, New York: J. J. Audubon, 1843–1844.

———. *The Birds of America, from Drawings Made in the United States and Their Territories.* 7 vols. New York: V. G. Audubon, 1856.

———. *The Birds of America, from Drawings Made in the United States and Their Territories.* 7 vols. New York: V. G. Audubon, 1859.

———. *The Birds of America, from Drawings Made in the United States and Their Territories.* 7 vols. New York: V. G. Audubon, Roe Lockwood and Son, 1860.

———. *The Birds of America, from Drawings Made in the United States and Their Territories.* 7 vols. New York: J. W. Audubon, Roe Lockwood and Son, 1861.

———. *The Birds of America, from Drawings Made in the United States and Their Territories.* 8 vols. New York: J. W. Audubon, Roe Lockwood & Son, 1865.

———. *The Birds of America, from Drawings Made in the United States and Their Territories.* 8 vols. New York: George R. Lockwood, [1870].

———. *The Birds of America, from Drawings Made in the United States and Their Territories.* 8 vols. New York: George R. Lockwood, [1871].

———. *The Birds of America, from Drawings Made in the United States and Their Territories.* 8 vols. New York: George R. Lockwood, 1889.

———. *The Birds of America, from Drawings Made in the United States and Their Territories.* 7 vols. Reissue of the 1840–1844 edition. New York: Dover Publications, 1967.

———. *The Birds of America, from Drawings Made in the United States and Their Territories.* 10 vols. Reissue. Kent, Ohio: Volair Limited, 1979. Also includes *The Quadrupeds of North America*.

———. *The Birds of America, from Drawings Made in the United States and Their Territories* and *The Quadrupeds of North America*. Sales sample. American Antiquarian Society, Worcester, Mass.

———. *The Birds of America, from Original Drawings.* London: J. J. Audubon, 1827–1838. Copy in the Stark Museum of Art, Orange, Texas.

———. *The Birds of America, from Original Drawings.* New York: Roe Lockwood and Son, 1860.

———. *The Imperial Collection of Audubon Animals: The Quadrupeds of North America.* Ed. by Victor H. Cahalane. New York: Bonanza Books, 1967.

———. *Multimedia Birds of America: A Replica of the Complete Works of John James Audubon's Birds of America (1840–1844).* n. p.: CMC Research, Inc., 1990. [CD-ROM]

———. *My Style of Drawing Birds.* Introduction by Michael Zinman. Ardsley, N.Y.: Overland Press, 1979.

———. *The Original Water-Color Paintings by John James Audubon for The Birds of America.* Introduction by Marshall B. Davidson. 2 vols. New York: American Heritage, 1966.

———. *Ornithological Biography, or an Account of the Habits of the Birds of the United States of America.* 5 vols. Edinburgh: Adam Black and Charles Black, 1831–1839.

———. *A Synopsis of the Birds of North America.* Edinburgh: Adam and Charles Black; London: Longman, Rees, Brown, Green and Longman, 1839.

———. *Three Letters of John James Audubon to John Stevens Henslow.* San Francisco: Grabhorn Press for the Members of the Roxburghe Club, 1943.

———. *The Viviparous Quadrupeds of North America.* 2 vols. New York: J. J. Audubon, 1845–1846.

Bibliography

———. *The Viviparous Quadrupeds of North America*. 3 vols. New York: J. J. Audubon and V. G. Audubon, 1846–1854.

———. *The Quadrupeds of North America*. 3 vols. New York: V. G. Audubon, 1854.

Audubon, John Woodhouse. *Illustrated Notes of an Expedition Through Mexico and California*. New York: J. W. Audubon, 1852.

Audubon, Lucy, ed. *The Life of John James Audubon, the Naturalist*. Introduction by James Grant Wilson. New York: G. P. Putnam and Son, 1869.

Audubon, Maria R., ed. *Audubon and His Journals*. Zoological and other notes by Elliott Coues. 2 vols. New York: Charles Scribner's Sons, 1897.

Beach, Moses Yale. *Wealth and Biography of the Wealthy Citizens of New York City*. 6th ed. New York: Sun Office, 1845.

Berlandier, Jean Louis. *Journey to Mexico during the Years 1826 to 1834*. 2 vols. Austin: Texas State Historical Association, 1980.

Berlandier, Jean Louis, and Rafael Chovel. *Diario de viage de la Comisión de Límites*. Mexico City: Juan R. Navarro, 1850.

Bonaparte, Charles Lucien. *American Ornithology; or, The Natural History of Birds Inhabiting the United States, Not Given by Wilson*. 4 vols. Philadelphia: Samuel Augustus Mitchell; Carey, Lea and Carey, 1825–1833.

Brotherhead, William. *Forty Years among the Old Booksellers of Philadelphia, with Bibliographical Remarks*. Philadelphia: A. P. Brotherhead, 1891.

Bryant, William Cullen. "A Forest Hymn." In George Perkins, Sculley Bradley, Richmond Croom Beatty, and E. Hudson Long, eds., *The American Tradition in Literature*. 6th ed. New York: Random House, 1985.

Cassin, John. *Illustrations of the Birds of California, Texas, Oregon, British and Russian America*. Philadelphia: J. B. Lippincott and Co., 1856. Reprint. Introduction by Robert McCracken Peck. Austin: Texas State Historical Association, 1991.

Catlin, George. *Letters and Notes on the Manners, Customs, and Conditions of North American Indians*. New York: Dover, 1973 reprint.

Christie's International Ltd. *Valuable Travel and Natural History Books*. London: Christie's, 1990.

Cincinnati Daily Gazette. 1833.

Corning, Howard, ed. *Journal of John James Audubon, Made during His Trip to New Orleans in 1820–1821*. Cambridge: Business Historical Society, 1939.

———, ed. *Journal of John James Audubon, Made While Obtaining Subscriptions to His Birds of America, 1840–1843*. Introduction by Francis H. Herrick. Boston: Club of Odd Volumes, 1929.

———, ed. *Letters of John James Audubon, 1826–1840*. 2 vols. Boston: Club of Odd Volumes, 1930.

Davidson, Marshall B., ed. *The Original Water-Color Paintings by John James Audubon for The Birds of America*. Introduction by Davidson. 2 vols. New York: American Heritage, 1966.

Deane, Ruthven, ed. "A Hitherto Unpublished Letter of John James Audubon." *Auk* 22 (April 1905): 170–175.

———, ed. "Some Letters of Bachman to Audubon." *Auk* 46 (April 1929): 177–184.

———, ed. "An Unpublished Letter of John James Audubon to His Family." *Auk* 25 (April 1908): 166–173.

———, ed. "Unpublished Letters of John James Audubon and Spencer F. Baird." *Auk* 21 (April 1904): 255–259.

———, ed. "Unpublished Letters of John James Audubon and Spencer F. Baird." *Auk* 23 (April and July 1906): 194–209, 318–334; 24 (January 1907): 53–70.

Dentzel, Carl Shaefer, ed. *The Drawings of John Woodhouse Audubon, Illustrating His Adventures Through Mexico and California, 1849–1850*. Introduction and notes by Shaefer. San Francisco: Book Club of California, 1957.

Dickson, Harold Edward, ed. *Observations on American Art: Selections from the Writings of John Neal (1793–1876)*. Pennsylvania State College Bulletin no. 12. State College, Pa., 1943.

Dunlap, William. *History of the Rise and Progress of the Arts of Design in the United States*. Introduction by William P. Campbell. Edited by Alexander Wyckoff. 3 vols. New York: Benjamin Blom, 1965.

Dwight, Edward H. "The Autobiographical Writings of John James Audubon." *Bulletin of the Missouri Historical Society* 19 (October 1962): 26–35.

Emory, William H. *Report on the United States and Mexican Boundary Survey*. 34th Cong., 1st sess., 1857. H.E.D. 135, S.E.D. 108. 2 vols. in 3 parts.

Ewers, John C., ed. *The Indians of Texas in 1830 by Jean Louis Berlandier*. Translated by Patricia Reading Leclercq. Washington, D.C.: Smithsonian Institution Press, 1969.

Fairchild, Herman Leroy. *History of the New York Academy of Sciences*. New York: Herman L. Fairchild, 1887.

Ford, Alice, ed. *Audubon, by Himself, A Profile of John James Audubon*. Garden City, N.Y.: Doubleday, 1969.

———, ed. *The 1826 Journal of John James Audubon*. Norman: University of Oklahoma Press, 1967.

Francis, C. S., and Co., New York. Advertisement for the "Works of John James Audubon," in possession of Hill Memorial Library, Louisiana State University, Baton Rouge.

Gibson's Guide and Directory of the State of Louisiana, and the Cities of New Orleans and Lafayette. New Orleans: John Gibson, 1838.

Godwin, Parke. "Audubon." In *Homes of American Authors; Comprising Anecdotical [sic] Personal, and Descriptive Sketches*. New York: G. P. Putnam and Co., 1853. 3–17.

———. *Commemorative Addresses: George William Curtis, Edwin Booth, Louis Kossuth, John James Audubon, William Cullen Bryant*. New York: Harper and Brothers, 1895.

———. "John James Audubon." *United States Magazine and Democratic Review* 10 (May 1842): 436–450.

Grinnell, George Bird. "Some Audubon Letters." *Auk* 33 (April 1916): 119–130.

Hammond, E. A. "Dr. Stroebel's Account of John J. Audubon." *Auk* 80 (1963): 464–466.

Holley, Mary Austin. *Texas*. Lexington, Ky.: J. Clarke and Co., 1836.

Hooton, Charles. *St. Louis' Isle, or Texiana*. London: Simmonds and Ward, 1847.

Houstoun, Matilda Charlotte (Jesse) Fraser. *Texas and the Gulf of Mexico; or Yachting in the New World*. 2 vols. London: John Murray, 1844.

James, Edwin, comp. *Account of an Expedition from Pittsburgh to the Rocky Mountains, under the Command of Major Stephen H. Long*. Introduction by Howard R. Lamar. Barre, Mass.: Imprint Society, 1972.

Lownes, Albert E. "Ten Audubon Letters." *Auk* 52 (April 1935): 154–168.

McDermott, John Francis, comp. and ed. *Audubon in the West*. Norman: University of Oklahoma Press, 1965.

———, ed. *Up the Missouri with Audubon: The Journal of Edward Harris*. Norman: University of Oklahoma Press, 1951.

Bibliography

Marcy, Randolph B. *Exploration of the Red River of Louisiana, in the Year 1852.* 33d Cong., 1st sess., 1854. S.E.D. 54.

Muir, Andrew Forest, ed. *Texas in 1837: An Anonymous, Contemporary Narrative.* Austin: University of Texas Press, 1958.

National Gazette and Literary Register (Philadelphia). 1832.

New Orleans Annual and Commercial Directory, 1844, Containing the Names and Residences of All the Inhabitants of the City and Suburbs of New Orleans. New Orleans: J. J. Calberthwaite and Co., 1845.

New Orleans Annual and Commercial Register for 1846. New Orleans: E. A. Michel and Co., 184—?

New Orleans Directory for 1841; Made by the United States Deputy Marshals (While Taking the Late Census), Containing the Names, Professions and Residences of All the Inhabitants of the City and Suburbs of New Orleans, Lafayette and Algiers. New Orleans: E. A. Michel and Co., 1840.

New Orleans Directory for 1842, Comprising the Names, Residences and Occupations of the Merchants, Business Men, Professional Gentlemen and Citizens of New Orleans, Lafayette, Algiers, and Gretna. New Orleans: Pitts and Clarke, 1842.

"Notice Concerning the Late Mr. Drummond's Journeys and His Collections, Made Chiefly in the Southern and Western Parts of the United States," *Companion to the Botanical Magazine* (London). 1 (1835): 39–46.

"Notice of Audubon's *Birds of America*." *American Journal of Science and Arts* 42 (April 1842): 130–136.

Northern Standard (Clarksville, Texas). 1846.

Peck, Robert McCracken. Introduction to *Illustrations of the Birds of California, Texas, Oregon, British and Russian America* by John Cassin. Reprint. Austin: Texas State Historical Association, 1991. I-1-38.

Publishers Weekly. 1906.

"Report . . . to Examine Splendid Work of Mr. Audubon," *Silliman's Journal* 16 (1829): 353–354.

Rhoads, Samuel N. "Auduboniana." *Auk* 20 (October 1903): 377–383.

———. "More Light on Audubon's Folio 'Birds of America,'" *Auk* 33 (April 1916): 130–132.

San Antonio Zeitung. 1855.

Saturday Courier (Philadelphia). 1840–1844.

Schoolcraft, Henry R. *Historical and Statistical Information Respecting the History, Condition and Prospects of the Indian Tribes of the United States.* 6 vols. Philadelphia: J. B. Lippincott and Co., 1851–1857.

Scudder, H. E., ed. *Recollections of Samuel Breck with Passages from His Note-Books (1771–1862).* Philadelphia: Porter and Coates, 1877.

Senefelder, Alois. *A Complete Course of Lithography.* Introduction by A. Hyatt Mayor. New York: Da Capo Press, 1977.

Southern Cabinet (Richmond). Vol. 1 (1840).

Squier, Ephraim George. *Ancient Monuments of the Mississippi Valley, Comprising the Results of Extensive Original Surveys and Explorations.* Smithsonian Contributions to Knowledge, vol. 1. Washington, D.C.: Smithsonian Institution, 1848.

Telegraph and Texas Register (Houston). 1837.

Thayer, John E. "Auduboniana." *Auk* 33 (April 1916): 115–118.

Torrey, Bradford, and Francis H. Allen, eds. *The Journals of Henry D. Thoreau.* 14 vols. Boston: Houghton Mifflin Co., 1949.

Townsend, John Kirk. *Narrative of a Journey across the Rocky Mountains, to the Columbia River, and a Visit to the Sandwich Islands, Chili, etc.* Philadelphia: Henry Perkins; Boston: Perkins and Marvin, 1839.

———. *Ornithology of the United States of America; or, Descriptions of the Birds Inhabiting the States and Territories of the Union.* Philadelphia: J. B. Chevalier, 1839.

———. *Sporting Excursions in the Rocky Mountains, Including a Journey to the Columbia River, and a Visit to the Sandwich Islands, Chili, etc.* London: Henry Colburn, 1840.

Tuckerman, Bayard, ed. *The Diary of Philip Hone, 1828–1851.* 2 vols. New York: Dodd, Mead and Co., 1889.

Tuckerman, Henry T. "Birds and Audubon." *Methodist Quarterly Review* 34 (July 1852): 415–428.

———. *Book of the Artists: American Artist Life Comprising Biographical and Critical Sketches of American Artists, Preceded by an Historical Account of the Rise and Progress of Art in America.* New York: James F. Carr, 1966.

Warren, B. H. *Report on the Birds of Pennsylvania, with Special Reference to the Food-Habits, Based on over Four Thousand Stomach Examinations.* 2d ed. Harrisburg, Pa.: E. K. Meyers, 1890.

Webber, Charles W. "About Birds and Audubon," *American Review: A Whig Journal* 2 (1845): 279–287.

———. *The Hunter-Naturalist: Romance of Sporting; or, Wild Scenes and Wild Hunters.* Philadelphia: J. W. Bradley, 1851; Lippincott, Grambo and Co., 1852.

———. *The Hunter-Naturalist: Wild Scenes and Song-Birds.* New York: G. P. Putnam and Co., 1854; Riker, Thorne and Co., 1854.

Winterfield, Charles. "A Talk about Birds." *American Review: A Whig Journal* 2 (September 1845): 279–287.

———. "American Ornithology." *American Review: A Whig Journal* 1 (March 1845): 262–275.

SECONDARY SOURCES

Adams, Alexander B. *John James Audubon: A Biography.* New York: G. P. Putnam's Sons, 1966.

Alexander, Edward P. *Museum Masters: Their Museums and Their Influence.* Nashville: American Association for State and Local History, 1983.

Allen, Elsa Guerdrum. "The History of American Ornithology before Audubon." *Transactions of the American Philosophical Society* 41 (1951): 385–591.

Arthur, Stanley Clisby. *An Intimate Life of the American Woodsman.* New Orleans: Harmanson, 1937.

———, ed. and comp. *Old Families of Louisiana.* Baton Rouge: Claitor's, 1971 reprint.

Audubon Centennial Exhibition: Souvenir Program. Reading, Pa.: Reading Public Museum and Art Gallery, 1951.

Bannon, Lois Elmer, and Taylor Clark. *Handbook of Audubon Prints.* Gretna, La.: Pelican, 1985.

Barber, Lynn. *The Heyday of Natural History, 1820–1870.* London: Jonathan Cape, 1980.

Barnhill, Georgia B. "The Publication of Illustrated Natural Histories in Philadelphia, 1800–1850." In Gerald W. R. Ward, *The American Illustrated Book in the Nineteenth Century.* Winterthur, Del.: Henry Francis du Pont Winterthur Museum, 1987; distributed by the University Press of Virginia, Charlottesville. 53–88.

Bell, James B. *John James Audubon: A Selection of Watercolors in the New-York Historical Society.* New York: New-York Historical Society, 1985.

Bibliography

Bennett, Whitman. *A Practical Guide to American Nineteenth Century Color Plate Books*. Rev. ed. Ardsley, N.Y.: Haydn Foundation for the Cultural Arts, 1980.

Brown, Milton W. *American Art: Painting, Sculpture, Architecture, Decorative Arts, Photography*. New York: Prentice-Hall, 1979.

Buchanan, Robert, ed. *The Life and Adventures of John James Audubon, the Naturalist, Edited by Robert Buchanan from Materials Supplied by His Widow*. London: Sampson Low, Son, and Marston, 1868.

Buff, Sheila, ed. *Audubon's Birds of North America*. Introduction by Buff. Stamford, Conn.: Longmeadow Press, 1990.

Cadbury, Warder H. "Alfred Jacob Miller's Chromolithographs." In Ron Tyler, ed., *Alfred Jacob Miller: Artist on the Oregon Trail*. Fort Worth: Amon Carter Museum, 1982. 447–449.

Cahalane, Victor H., ed. *The Imperial Collection of Audubon Animals: The Quadrupeds of North America*. New York: Bonanza Books, 1967.

Cantwell, Robert. *Alexander Wilson, Naturalist and Pioneer*. Philadelphia and New York: J. B. Lippincott and Co., 1961.

Catalogue of the John James Audubon Memorial Museum. Henderson, Ky.: Kentucky State Parks, n.d.

Cluck, Nancy. "Audubon: Images of the Artist in Eudora Welty and Robert Penn Warren." *Southern Literary Journal* 17 (Spring 1985): 41–53.

Coffin, Annie Roulhac. "Audubon's Friend—Maria Martin." *New-York Historical Society Quarterly* 49 (January 1965): 28–51.

Conrad, Glenn R., ed. *A Dictionary of Louisiana Biography*. 2 vols. Lafayette: Louisiana Historical Association in cooperation with the Center for Louisiana Studies, University of Southwestern Louisiana, 1988.

Contemplating the American Watercolor: Selections from the Transco Energy Company Collection. Houston: N.p., n.d.

Cosentino, Andrew F. *The Paintings of Charles Bird King (1785–1862)*. Washington, D.C.: National Collection of Fine Arts, Smithsonian Institution Press, 1977.

Coues, Elliott. *Birds of the Colorado Valley, a Repository of Scientific and Popular Information Concerning North American Ornithology*. Miscellaneous Publications No. 11. Washington, D.C.: Department of the Interior, U.S. Geological Survey of the Territories, 1878.

Cowdrey, Mary Bartlett. *National Academy of Design of Exhibition Records, 1826–1860*. 2 vols. New York: New-York Historical Society, 1943.

Dance, S. Peter. *The Art of Natural History: Animal Illustrators and Their Work*. Woodstock, N.Y.: Overlook Press, 1978.

Darnton, Robert. *The Great Cat Massacre and Other Episodes in French Cultural History*. New York: Vintage Books, 1985.

———. "What Is the History of Books?" *Daedalus* 3 (Summer 1982): 65–83.

Delatte, Carolyn E. *Lucy Audubon, a Biography*. Baton Rouge and London: Louisiana State University Press, 1982.

Dippie, Brian W. *Catlin and His Contemporaries: The Politics of Patronage*. Lincoln: University of Nebraska Press, 1990.

Dobie, J. Frank. *Rattlesnakes*. Boston: Little, Brown, 1965.

Dock, George, Jr. *Audubon's Birds of America*. New York: Arrowood Press, 1987.

Dormon, James H., ed. *Audubon: A Retrospective*. Lafayette, La.: Center for Louisiana Studies, University of Southwestern Louisiana, 1990.

Bibliography

Doughty, Robin W. *The Mockingbird*. Austin: University of Texas Press, 1988.

———. *Return of the Whooping Crane*. Austin: University of Texas Press, 1989.

Dunning, B. "An Unsung Family of Engravers: The Havells of Berkshire." *Country Life* 153 (February 22, 1973): 458–459.

Durant, Mary, and Michael Harwood. "In Search of the Real Mr. Audubon," *Audubon* 87 (May 19, 1985): 58–115.

———. *On the Road with John James Audubon*. New York: Dodd, Mead and Co., 1980.

Durrell, Jane. "Upstairs at Mrs. Amelung's." *Cincinnati Historical Society Bulletin* 35 (Spring 1977): 32–41.

Dwight, Edward H. "Audubon in Kentucky." *Antiques* 105 (April 1974): 850–854.

———. *Audubon Watercolors and Drawings*. Utica, N.Y.: Munson-Williams-Proctor Institute; New York: Pierpont Morgan Library, 1965.

Ewers, John C. "Artists' Choices." *American Indian Art Magazine* 7 (Spring 1982): 40–49.

Feduccia, Alan, ed. *Catesby's Birds of Colonial America*. Chapel Hill and London: University of North Carolina Press, 1985.

Ferber, Linda S., and William H. Gerdts. *The New Path: Ruskin and the American Pre-Raphaelites*. Brooklyn: Brooklyn Museum, 1989.

Field, Richard S. "Audubon's Lithograph of the Clapper Rail." *Art Quarterly* 29 (Spring 1966): 147–153.

Fiero, Gloria K. "Audubon the Artist." In James H. Dorman, ed., *Audubon: A Retrospective*. Lafayette, La.: Center for Louisiana Studies, University of Southwestern Louisiana, 1990. 34–60.

Flexner, James Thomas. *America's Old Masters: First Artists of the New World*. New York: Viking Press, 1939.

———. *History of American Painting, 1760–1835: The Light of Distant Skies*. 2 vols. New York: Dover, 1969.

———. *That Wilder Image: The Painting of America's Native School from Thomas Cole to Winslow Homer*. New York: Bonanza Books, 1962.

Flint, Janet. "The American Painter-Lithographer." In *Art and Commerce: American Prints of the Nineteenth Century*. Boston: Museum of Fine Arts, 1978. 126–142.

Ford, Alice, comp. and ed. *Audubon's Animals: The Quadrupeds of North America*. New York: The Studio Publications, Inc., in association with Thomas Y. Crowell Company, 1951.

———, comp. and ed. *Audubon's Butterflies, Moths, and Other Studies*. New York: The Studio Publications, Inc., in association with Thomas Y. Crowell Company, 1952.

———. *John James Audubon*. Norman: University of Oklahoma Press, 1964.

———. *John James Audubon: A Biography*. New York: Abbeville Press, 1988.

Foshey, Ella M. *Reflections of Nature: Flowers in American Art*. New York: Alfred A. Knopf, 1984.

Frantz, Joe B. *Gail Borden: Dairyman to a Nation*. Norman: University of Oklahoma Press, 1951.

Fries, Waldemar H. *The Double Elephant Folio: The Story of Audubon's Birds of America*. Chicago: American Library Association, 1973.

———. "Joseph Bartholomew Kidd and the Oil Paintings of Audubon's *Birds of America*." *Art Quarterly* 26 (Autumn 1967): 339–346.

Gascoigne, Bamber. *How to Identify Prints: A Complete Guide to Manual and Mechanical Processes From Woodcut to Ink Jet*. New York: Thames and Hudson, 1986.

Geikie, Archibald. *The Founders of Geology*. 2d ed. London: Macmillan and Co., 1905.

Bibliography

Geiser, Samuel Wood. "Naturalist of the Frontier: Audubon in Texas." *Southwest Review* 16 (Autumn 1930): 109–135.

———. *Naturalists on the Frontier.* 2d ed. Dallas: Southern Methodist University Press, 1948.

———. "Thomas Drummond." *Southwest Review* 15 (Summer 1930): 478–512.

Gentling, Scott, John Graves, and Stuart Gentling. *Of Birds and Texas.* Fort Worth: Gentling Editions, 1986.

Gifford, George E., and Florence B. Smallwood. "Sidelights: Audubon's 'View of Baltimore,'" *Maryland Historical Magazine* 72 (Summer 1977): 266–271.

Gilreath, James. "American Book Distribution." In David D. Hall and John B. Hench, eds., *Needs and Opportunities in the History of the Book: America, 1639–1876.* Worcester, Mass.: American Antiquarian Society, 1987. 103–185.

Glanz, Dawn. *How the West Was Drawn: American Art and the Settling of the Frontier.* Ann Arbor: UMI Research Press, 1982.

Goetzmann, William H. *Exploration and Empire: The Explorer and the Scientist in the Winning of the American West.* New York: Alfred A. Knopf, 1967.

Goetzmann, William H., and William N. Goetzmann. *The West of the Imagination.* New York: W. W. Norton and Co., 1986.

Gopnik, Adam. "Audubon's Passion." *New Yorker* 67 (February 25, 1991): 96–104.

Graustein, Jeannette E. *Thomas Nuttall, Naturalist: Explorations in America, 1808–1841.* Cambridge, Mass.: Harvard University Press, 1967.

Hall, David D., and John B. Hench, eds. *Needs and Opportunities in the History of the Book: America, 1639–1876.* Worcester, Mass.: American Antiquarian Society, 1987.

Hammond, John H., and Jill Austin. *The Camera Lucida in Art and Science.* Bristol, England: Adam Hilger, 1987.

Hart, Mary Bell. *Audubon's Texas Quadrupeds: A Portfolio of Color Prints.* Austin: Hart Graphics, 1979.

Harwood, Michael. *Audubon Demythologized.* New York: National Audubon Society, n.d.

———. "Mr. Audubon's Last Hurrah," *Audubon* 87 (November 18, 1985), 80–116.

Hendricks, Gordon. *The Life and Work of Winslow Homer.* New York: Harry N. Abrams, 1979.

Herrick, Francis Hobart. *Audubon the Naturalist: A History of His Life and Time.* 2 vols. New York and London: D. Appleton and Co., 1917.

———. "Audubon's Bibliography." *Auk* 37 (July 1919): 372–380.

Hill, Jim Dan. *The Texas Navy: In Forgotten Battles and Shirtsleeve Diplomacy.* Austin: State House Press, 1987.

Howat, John K. Introduction to *American Paradise: The World of the Hudson River School.* New York: Metropolitan Museum of Art, 1987.

Hummer, T. R. "Robert Penn Warren: Audubon and the Moral Center." *Southern Review* 16 (1980): 799–814.

Huntington, David Carew. *Art and the Excited Spirit: America in the Romantic Period.* Ann Arbor: University of Michigan Museum of Art, 1972.

Huth, Hans. *Nature and the American: Three Centuries of Changing Attitudes.* Lincoln and London: University of Nebraska Press, 1972.

Irwin, Francina. "The Man in the Wolfskin Coat: John James Audubon in Edinburgh (1826–27)." *Country Life* 161 (April 28, 1977): 1104–1106.

Bibliography

Jackson, Christine E. *Bird Etchings: The Illustrators and Their Books, 1655–1855.* Ithaca and London: Cornell University Press, 1985.

———. *Bird Illustrators: Some Artists in Early Lithography.* London: H. F. and G. Witherby, 1975.

Jackson, Donald. *Voyages of the Steamboat Yellow Stone.* New York: Ticknor and Fields, 1985.

Jenkins, John H. *Audubon and Texas.* Austin: Pemberton Press, 1965.

Johns, Elizabeth. *American Genre Painting: The Politics of Everyday Life.* New Haven: Yale University Press, 1991.

Jordan, Terry G. *German Seed in Texas Soil: Immigrant Farmers in Nineteenth-Century Texas.* Austin: University of Texas Press, 1966.

Jussim, Estelle. *Frederic Remington, the Camera and the Old West.* Fort Worth: Amon Carter Museum, 1983.

Keating, L. Clark. *Audubon, the Kentucky Years.* Lexington: University Press of Kentucky, 1976.

Kennedy, Caroline. *Building and Planning for the Future: The E. A. McIlhenny Natural History Collection at Louisiana State University, 1971–1991: An Exhibition Celebrating the Twentieth Anniversary of the Collection.* Introduction by David M. Lank. Baton Rouge: LSU Libraries, Special Collections, 1991.

Kolodny, Annette. *The Lay of the Land: Metaphor As Experience and History in American Life and Letters.* Chapel Hill: University of North Carolina Press, 1975.

Ladner, Mildred D. "Artists on the Missouri." *Southwest Art* 12 (October 1982): 117–123.

Lambourne, Maureen. "Romanticism or Realism? The Bird Paintings of Audubon and Gould." *Country Life* 175 (January 26, 1984): 232–233.

Lane, Mills. *Architecture of the Old South: Louisiana.* New York: Abbeville Press, 1990.

Larkin, Jack. *The Reshaping of Everyday Life, 1790–1840.* New York: Harper and Row, 1988.

Larkin, Oliver W. *Art and Life in America.* New York: Rinehart, 1949.

Lindsey, Alton A. *The Bicentennial of John James Audubon.* Bloomington: University of Indiana Press, 1985.

Low, Susanne M. *An Index and Guide to Audubon's Birds of America: A Study of the Double Elephant Folio of John James Audubon's Birds of America, as Engraved by William H. Lizars and Robert Havell.* New York: American Museum of Natural History, Abbeville Press, 1988.

McComb, David G. *Houston: A History.* Rev. ed. Austin: University of Texas Press, 1981.

McCoubrey, John W. *American Tradition in Painting.* New York: George Braziller, 1963.

McCusker, John J. "How Much Is That in Real Money? A Historical Price Index for Use As a Deflator of Money Values in the Economy of the United States." *Proceedings of the American Antiquarian Society: A Journal of American History and Culture Through 1876* 101 (Pt. 2, 1992): 297–373.

McEwen, John. "Audubon and His Legacy." *Art in America* 73 (September 1985): 99–108.

McGrath, Daniel Francis. "American Colorplate Books, 1800–1900." Ph.D. diss., University of Michigan, 1966.

Martin, Fontaine. *A History of the Bouligny Family and Allied Families.* Lafayette: Center for Louisiana Studies, University of Southwestern Louisiana, 1990.

Marzio, Peter H. "American Lithographic Technology before the Civil War." In John D. Morse, ed., *Prints in and of America to 1850: Sixteenth Annual Winterthur Conference, 1970.* Charlottesville: University Press of Virginia, 1970. 215–256.

———. *The Democratic Art: Chromolithography, 1840–1900: Pictures for a 19th-Century America.* Boston: David R. Godine, 1979.

Bibliography

———. "The Democratic Art of Chromolithography in America: An Overview." In *Art and Commerce: American Prints of the Nineteenth Century*. Boston: Museum of Fine Arts, 1978. 76–102.

———. *Mr. Audubon and Mr. Bien: An Early Phase in the History of American Chromolithography*. Washington, D.C.: National Museum of History and Technology, Smithsonian Institution, 1975.

Meador, Shirley. "John James Audubon: Artist and Naturalist." *American Artist* 38 (May 1974): 53–54, 57, 72–75.

Merrill, Lynn L. *The Romance of Victorian Natural History*. New York: Oxford University Press, 1989.

Meyers, Army R. W. "Imposing Order on the Wilderness: Natural History Illustration and Landscape Portrayal." In Edward J. Nygren, ed., *Views and Visions: American Landscape before 1830*. Washington, D.C.: Corcoran Gallery of Art, 1986. 105–131.

Mitchell, Lee Clark. *Witness to a Vanishing America: The Nineteenth-Century Response*. Princeton, N.J.: Princeton University Press, 1981.

Morgan, Ann Lee. "The American Audubons: Julius Bien's Lithographed Edition." *Print Quarterly* 4 (1987): 362–379.

Morton, Ohland. *Terán and Texas: A Chapter in Texas-Mexican Relations*. Austin: Texas State Historical Association, 1948.

Munsing, Stefanie A. *Made in America: Printmaking, 1760–1860: An Exhibition of Original Prints from the Collections of the Library Company of Philadelphia and the Historical Society of Pennsylvania, April–June, 1973*. Philadelphia: Library Company of Philadelphia, 1973.

Newhall, Beaumont. *The History of Photography from 1839 to the Present*. Rev. ed. New York: Museum of Modern Art, 1982.

Novak, Barbara. *American Painting of the Nineteenth Century: Realism, Idealism, and the American Experience*. New York: Praeger Publishers, 1969.

———. *Nature and Culture: American Landscape and Painting, 1825–1875*. New York: Oxford University Press, 1980.

Orosz, Joel J. *Curators and Culture: The Museum Movement in America, 1740–1870*. Tuscaloosa: University of Alabama Press, 1990.

Owens, Carlotta J. *John James Audubon, The Birds of America*. Washington, D.C.: National Gallery of Art, 1985.

Parry, Elwood C. "Thomas Cole's Early Career: 1818–1829." In Edward J. Nygren, ed., *Views and Visions: American Landscape before 1830*. Washington, D.C.: Corcoran Gallery of Art, 1986. 160–187.

Peattie, Donald C. *Green Laurels: The Lives and Achievements of the Great Naturalists*. New York: Simon and Schuster, 1935.

———, ed. *Audubon's America: The Narratives and Experiences of John James Audubon*. Boston: Houghton Mifflin Co., 1940.

Perkins, George Sculley Bradley, Richmond Croom Beatty, and E. Hudson Long, eds. *The American Tradition in Literature*. 6th ed. New York: Random House, 1985.

Perrault, Anna H., comp. and ed. *Nature Classics: A Catalogue of the E. A. McIlhenny Natural History Collection at Louisiana State University*. Preface by Kathryn Morgan. Introduction by David M. Lank. Baton Rouge: Friends of the LSU Library, 1987.

Pessen, Edward. *Jacksonian America: Society, Personality, and Politics*. Homewood, Ill.: Dorsey Press, 1969.

Peters, Harry T. *Currier and Ives: Printmakers to the American People*. Garden City, N.Y.: Doubleday, Doran and Co., 1942.

Bibliography

Peterson, Roger Tory, ed. *The Art of Audubon: The Complete Birds and Mammals: John James Audubon.* Introduction by Peterson. New York: Times Books, 1979.

———. "John James Audubon, 1785–1851." *Southwest Art* 12 (November 1982): 104, 106–111.

Peterson, Roger Tory, and Virginia Marie Peterson, eds. *Audubon's Birds of America: The Audubon Society Baby Elephant Folio.* Rev. ed. New York: Abbeville Press, 1990.

Pickens, Donald K. "Handmaidens to History." *Reviews in American History* 11 (September 1983): 351–355.

Pinckney, Pauline A. *Painting in Texas: The Nineteenth Century.* Austin: University of Texas Press, 1967.

Powell, Earl A. *Thomas Cole.* New York: Harry N. Abrams, 1990.

Proby, Kathryn Hale. *Audubon in Florida, with Selections from the Writings of John James Audubon.* Coral Gables, Fla.: University of Miami Press, 1974.

Reese, William S. "The Bonaparte Audubons at the Amon Carter Museum and the Friendship of John James Audubon and Charles Lucien Bonaparte." In Ron Tyler, ed., *Prints of the American West: Papers Presented at the Ninth Annual North American Print Conference.* Fort Worth: Amon Carter Museum, 1983. 13–24.

Remini, Robert V. *Andrew Jackson and the Course of American Democracy, 1833–1845.* Vol. 3. New York: Harper and Row, 1984.

Rewald, John. "Degas and His Family in New Orleans." In John Rewald, James B. Byrnes, and Jean Sutherland Boggs, *Edgar Degas, His Family and Friends in New Orleans.* New Orleans: Isaac Delgado Museum of Art, 1965.

Reynolds, Gary A. *John James Audubon and His Sons.* New York: Grey Art Gallery and Study Center, New York University, 1982.

Rice, Howard C., Jr., comp. "The World of John James Audubon: Catalogue of an Exhibition in the Princeton University Library, 15 May–30 September 1959." Introduction by Waldemar H. Fries. *Princeton University Library Chronicle* 21 (Autumn 1959/Winter 1960): 9–103.

Richards, Irving T. "Audubon, Joseph R. Mason, and John Neal." *American Literature* 6 (1934): 122–140.

Ripley, S. Dillon, and Lynette L. Scribner, comps. *Ornithological Books in the Yale University Library, Including the Library of William Robertson Coe.* New Haven: Yale University Press, 1961.

Rodriquez Roque, Oswaldo. "The Exaltation of American Landscape Painting." In John K. Howatt, ed., *American Paradise: The World of the Hudson River School.* New York: Metropolitan Museum of Art, 1987. 21–48.

Rosen, Charles, and Henri Zerner. *Romanticism and Realism: The Mythology of Nineteenth Century Art.* London: Faber and Faber, 1984.

Rosta, Paul. "John James Audubon." *American History Illustrated* 20 (October 1985): 22–31.

Rourke, Constance. *Audubon.* New York: Harcourt, Brace and Co., 1946.

Rudisill, Richard. *Mirror Image: The Influence of the Daguerreotype on American Society.* Albuquerque: University of New Mexico Press, 1971.

Sage, John H. "Description of Audubon." *Auk* 34 (1917): 239–240.

Sellers, Charles. *The Market Revolution: Jacksonian America, 1815–1846.* New York: Oxford University Press, 1991.

Sellers, Charles Coleman. *Mr. Peale's Museum: Charles Willson Peale and the First Popular Museum of Natural Science and Art.* New York: W. W. Norton and Co., 1980.

Bibliography

Shaw, Robert K. "Elihu Burritt–Friend of Mankind." *Proceedings of the American Antiquarian Society* (April 1926): 24–37.

Shelley, Donald A. "Audubon's Technique As Shown in His Drawings of Birds," *Antiques* 49 (1946): 354–357.

Siegel, Stanley. *A Political History of the Texas Republic, 1836–1845*. Austin: University of Texas Press, 1956.

Sinclair, Bruce. *Philadelphia's Philosopher Mechanics: A History of the Franklin Institute, 1824–1865*. Baltimore: Johns Hopkins University Press, 1974.

Sitwell, Sacheverell, Handasyde Buchanan, and James Fisher. *Fine Bird Books, 1700–1900*. London and New York: Collins and Van Nostrand, 1953.

Sotheby's. *The Library of H. Bradley Martin: John James Audubon, Magnificent Books and Manuscripts*. New York: Sotheby's, 1989.

———. *The Library of H. Bradley Martin: Magnificent Color-Plate Ornithology*. New York: Sotheby's, 1989.

Stapp, William F. "Daguerreotypes onto Stone: The Life and Work of Francis D'Avignon." In Wendy Wick Reaves, ed., *American Portrait Prints: Proceedings of the Tenth Annual American Print Conference*. Charlottesville: University Press of Virginia, for the National Portrait Gallery, Smithsonian Institution, 1984. 194–231.

Stebbins, Theodore E. *American Master Drawings and Watercolors: A History of Works on Paper from Colonial Times to the Present*. New York: Harper and Row, 1976.

———. "Luminism in Context: A New View." In John Wilmerding, *American Light: The Luminist Movement, 1850–1875: Paintings, Drawings, Photographs*. Washington, D.C.: National Gallery of Art, 1980. 211–234.

Stebbins, Theodore E., Carol Troyen, and Trevor J. Fairbrother. *A New World: Masterpieces of American Painting, 1760–1910*. Boston: Museum of Fine Arts, 1983.

Stewart, Rick, Joseph D. Ketner II, and Angela L. Miller. *Carl Wimar: Chronicler of the Missouri River Frontier*. Fort Worth: Amon Carter Museum, 1991.

Stone, Witmer. "A Bibliography and Nomenclator of the Ornithological Works of John James Audubon." *Auk* 23 (July 1906): 298–312.

Tate, Peter. *Birds, Men and Books: A Literary History of Ornithology*. London: Henry Sotheran Ltd., 1986.

Taylor, Joshua. *America As Art*. New York: Harper and Row, 1976.

"Texas Collection." *Southwestern Historical Quarterly* 49 (April 1946): 615–666.

Thomas, Keith. *Man and the Natural World: A History of the Modern Sensibility*. New York: Pantheon Books, 1983.

Thomas, Samuel W., and Eugene H. Conner. "John James Audubon and His Relationship with the Croghan Family of Louisville, Kentucky." *Register of the Kentucky Historical Society* 67 (1969): 237–247.

Tooley, R. V. *English Books with Coloured Plates, 1790 to 1860: A Bibliographical Account of the Most Important Books Illustrated by English Artists in Colour Aquatint and Colour Lithography*. London: B. T. Batsford, 1987.

Truettner, William H. *The Natural Man Observed: A Study of Catlin's Indian Gallery*. Washington, D.C.: Smithsonian Institution Press, 1979.

———, ed. *The West As America: Reinterpreting Images of the Frontier, 1820–1920*. Washington, D.C.: Smithsonian Institution Press, 1991.

Bibliography

Twyman, Michael. *Lithography, 1800–1850: The Techniques of Drawing on Stone in England and France and Their Application in Works of Topography*. London: Oxford University Press, 1970.

Tyler, Ron. *Nature's Classics: John James Audubon's Birds and Animals*. Orange, Texas: Stark Museum of Art, 1992.

———, ed. *Alfred Jacob Miller: Artist on the Oregon Trail*. Fort Worth: Amon Carter Museum, 1982.

Van Deusen, Glyndon G. *The Jacksonian Era, 1828–1848*. New York: Harper and Row, 1959.

Viola, Herman J. *Thomas L. McKenney, Architect of America's Early Indian Policy, 1816–1830*. Chicago: Swallow Press, 1974.

Wainwright, Nicholas B. *Philadelphia in the Romantic Age of Lithography*. Philadelphia: Historical Society of Pennsylvania, 1958.

Watters, Robinson C. "Audubon and His Baltimore Patrons." *Maryland Historical Magazine* 34 (June 1939): 138–143.

Webb, Max A. "*Audubon: A Vision:* Robert Penn Warren's Response to Eudora Welty's 'A Still Moment.'" *Mississippi Quarterly* 34 (1981): 445–455.

Welch, Margaret Curzon. "John James Audubon and His American Audience: Art, Science, and Nature, 1830–1860." Ph.D. diss., University of Pennsylvania, 1988.

Welker, Robert Henry. *Birds and Men: American Birds in Science, Art, Literature, and Conservation, 1800–1900*. Cambridge, Mass.: Harvard University Press, 1955.

Williams, George Alfred. "Portraits of Robert Havell, Junior, Engraver of Audubon's 'The Birds of America,'" *Print Collector's Quarterly* 7 (October 1917): 298–304.

———. "Robert Havell, Junior, Engraver of Audubon's *The Birds of America*." *Print Collector's Quarterly* 6 (1916): 227–257.

Wilmerding, John. *American Art*. New York: Penguin Books, 1976.

———. *American Light: The Luminist Movement, 1850–1875: Paintings, Drawings, Photographs*. Washington, D.C.: National Gallery of Art, 1980.

———. *Audubon, Homer, Whistler and 19th Century America*. New York: Lamplight, 1975.

———. "Winslow Homer's *Right and Left*." *Studies in the History of Art* 9 (1980): 59–85.

———, ed. *The Genius of American Painting*. New York: William Morrow and Co., 1973.

Zabriskie, George A. "The Story of a Priceless Art Treasure." *New-York Historical Society Quarterly* 30 (April 1946): 68–76.

Zimmer, John Todd. *Catalogue of the Edward E. Ayer Ornithological Library*, Pt. 1. Chicago: Field Museum of Natural History, Publication 239, Zoological Series, Vol. 16, 1926.

Index

Abbeville Press, 167n.2
Abert, Colonel John James, 70
Academy of Natural Sciences, 9, 109, 131, 169n.18
Ackerman, James T., 110, 178n.21
Albion (New York), 60, 140; comments of, regarding octavo *Birds*, 58, 69, 102, 104
Aldrovandi, Ulisse, 7
Allston, Washington, 150
Amadon, Dean, 165
American Academy of Arts and Sciences, 30
American Antiquarian Society, 71, 187n.36
American Avocet, 137
American Black or Silver Fox, 108
American Coot, 56, 62pl.23
American Flamingo, 39, 116
American Journal of Science and Arts, 52, 71
American Ornithology . . . (Bonaparte), 9, 10, 20pl.9, 169n.18
American Ornithology . . . (Wilson), 3, 8, 9, 13, 14, 21pl.11
American Philosophical Society, 31
American Redstart, 101
American Snipe, 138
Amon Carter Museum (Fort Worth), 25, 26, 171n.37, 177n.17
Analectic Magazine, 52
André, Mrs., 5, 169n.10
Arctic Fox, 108
Arctic Yager, 39
Arkansaw Flycatcher, 98pl.48, 99pl.49

Arthur's Magazine, 104
Assiniboins, 107
Audubon, Eliza Bachman (daughter-in-law), 74, 76, 78
Audubon, Jean (father), 3
Audubon, John James: appearance of, 12-13, 15, 16, 59, 149; article by, on rattlesnakes, 27; and Bowen, 80-81; and camera lucida, 66pl.29; in Charleston, 31; as competitor, 48-49; criticism of, 59-60; death of, 108; early employment of, 3, 4, 5; earnings of, 103-04, 183n.30; exhibitions of, 12, 13, 15, 16, 30, 48; and family, 130; finances of, 46, 82, 105; first return of, from England, 27-28; in Great Britain, 12-26, 35; and inclusion of habitats, 6, 15, 17pl.2, 37-39; involvement of, in production of octavo, 77-78; in Labrador, 31; in Louisiana, 6, 38; on the Missouri River, 84, 100pl.50, 106-07; in New York, 11, 27; and nomenclature, 109; and oil painting, 15-16, 132; in Paris, 26-27; payments to, 76, 87pl.34; in Philadelphia, 6, 8-11; portrait of, 149; qualities of paintings of, 15, 39-40, 76-77, 93pl.43; quest of, to draw all birds, 36, 39; and the Rocky Mountains, 31; as Romantic, 1, 32, 131-40, 148pl.67, 149-50; as salesman, 58-59, 60; in Scotland, 15; second return of, from England, 30, 131; self-portrait of, frontispiece, 13, 170n.26; senility of, 107; shortcuts of, 28, 39; techniques of, 5, 6, 15; in Texas, 36-38

Index

Audubon, John Woodhouse (son): as artist, 35, 74-75, 77, 89pl.36, 93pl.42, 123pl.58, 106, 108, 139, 186n.24; in California, 115; and camera lucida, 37, 51, 54; and Cassin, 111; and changes to plates, 115, 121pl.55, 125; death of, 128; as employee, 55, 76, 177n.20; and England, 28; finances of, 128; in Florida, 31; as manager, 115, 125, 128; marriage of, 38; in Texas, 37, 107; wife's illness, 70, 74-75

Audubon, Lucy Bakewell (wife), 3, 5, 28, 47, 48, 60; finances of, 128

Audubon, Maria Bachman (daughter-in-law), 38, 70, 74-75

Audubon Memorial Museum, 77

Audubon's Wood-Warbler, 82

Audubon, Victor Gifford (son): as business manager, 51, 55, 60, 70, 73, 75, 83, 106; and Cassin's proposal, 109, 111, 112; and changes to plates, 92pl.41, 121pl.55; in Cuba, 78; death of, 128; and double elephant folio, 33, 35; and England, 28; marriage of, 74; and painting, 35, 139; and royal octavo, 33-35, 93pl.42; spinal injury of, 115, 186n.23; and text, 107; and wife's illness, 76

A. V. (artist), 56, 163

Avedon, Richard (mention), 58

Bachman, Reverend John, 31, 36; as advisor, 36, 56-57, 102; as coauthor, 106, 107; and daughter of, 38; portrait of, 75; as reviewer, 57, 59; and selling, 61

Baird, Spencer F., 83, 102, 110, 111, 114

Bakewell, Thomas W. (brother-in-law), 4

Bakewell, William (father-in-law), 3

Bald eagle, 136

Barnet and Doolittle, 52

Barn Owl, 40, 116

Barred Owl, 28, 101

Barton, Benjamin Smith, 6

Bartram, John, 6, 138

Bartram's Vireo or Greenlet, 84

Bayou Sara, Louisiana, 6, 18pl.4, 38, 41pl.17

Bell, J. G., 84

Bell's Vireo, 77

Belon, Pierre, 7

Belted Kingfisher, 113, 116

Berlandier, Jean Luis, 36

Bewick, Thomas, 7, 16, 22pl.12, 26, 135

Beyer, Edward, 177n.17

Bien, Julius: history of, 125; and reprint of *Birds*, 125-28

Birchfield, Charles, 150

"Bird of Washington," 136

Birds of America, The (double elephant folio), 1-16, 25-32, 33-40, 45-46; coloring of, 29; composites in, 50, 64pl.25-26, 65pl.27-28; copperplates for, 47, 113, 125; cost of production of, 46; inclusion of eggs in, 171n.43; last proof of, 45; price of, 14, 46, 170n.29; 175n.25; print orders of, 14-15; profit of, 15, 183n.30; progress of, 35; prospectus for, 14, 24pl.16; reprints of, 125-28; sales of, 48; subscribers to, 14-15, 27, 30, 46, 70; survival of plates for, 128; text for, 28

Birds of America, The (royal octavo), 1, 2, 47-60, 69-72, 73-84, 101-05; coloring of, 62pl.23; comparisons among plates of, 81-82, 84, 101, 113-14, 115-16, 125, 129-30; completion of, 84, 102, 107; contents of, 84; copyright of, 73; cost of production of, 55, 69, 178nn.20, 21; 180n.37; cover of, 68pl.31; design of, 58, 69; differences in editions of, 113-16, 125-27, 129; drawings for, 74-76, 84, 101; and economy, 82-83; employees for, 79, 80; hybrid sets of, 79, 108; idea for, 26, 33-35; impact of, 2; oddities in, 79; partial set of, 185n.18; plates of, compared to double elephant, 89pl.36; popularity of, 82, 129; price of, 58, 176n.10; print orders of, 55, 69-70, 78, 101, 113; problems with, 73-75, 78-79, 82-83; profit of, 183n.30; publicity for, 60; reviews of, 104; scheme of, 51; size of edition of, 101; stones of, 129; subscribers to, 2, 58, 69-71, 75, 101-03; subsequent editions of, 2, 108, 112, 113, 115, 127, 129; text of, 101

Birds of Great Britain (Bewick), 22pl.12

Index

Birds of Pennsylvania. See Report on the Birds of Pennsylvania
Black-Bellied Plover, 77
Black-Billed Cuckoo, 6, 17pl.3, 26, 138
Black-shouldered Elanus, 79, 182n.20
Black, or Surf Duck, 39, 42pl.19
Black-Throated Green Warbler, 39
Black-throated Grey Wood Warbler, 82
Black-throated Guillemot, 40, 42pl.18, 139
Black-Throated Wax Wing, 75, 86pl.33
Black Vulture or Carrion Crow, 115
Blackwood's, 30
Blue Jay, 136, 142pl.61
Blue Mountain Warbler, 84
Boat-Tailed Grackles (in Bonaparte), 169n.18. *See also Great Crow Blackbird*
Bodmer, Karl, 8, 16, 107
Bohn, Henry G., 13, 26
Bonaparte, Charles-Lucien, 8, 9-10, 14, 20pl.9, 168n.4, 169n.18, 171n.37
Bonapartian Gull, 39
Boone, Daniel, 149, 179n.31
Boston Athenaeum, 30
Bowen and Company, 114
Bowen, John T.: and Cassin, 110, 111, 114; contributions of, to plates, 82, 113, 118pl.52; death of, 114; finances of, 74, 80, 182n.17; history of, 52; plates of, compared to Havell's, 81-82, 84, 101; problems with, 99pl.49; and production of royal octavo, 55-56, 70, 78; and *Quadrupeds*, 106; and simplification of drawings, 101; watercolor of, 177n.13.
Brady, Matthew, 148pl.67, 149
Brasilian Caracara Eagle, 136
Breck, Samuel, 48, 59
Brevis narratorio ... (De Bry), 7
Brewer, Thomas M., 38
British Copyright Act of 1709, 30
Broad-Winged Buzzard, 114
Brooklyn Museum, 164
Brooks, Joshua S., 13
Brotherhead, William, 127
Brother Jonathan, 59, 104

Brown, Captain Thomas, 33, 34
Brown, Milton W., 133
Brown thrashers, 136
Brown University, 164
Bruce, Mrs. (subscriber), 102
Bryant, William Cullen, 60, 140
Buffalo Bull's Back Fat, Head Chief, Blood Tribe (Catlin), 137, 144pl.63
Buff Breasted Sandpiper, 40
Bufford, John H., 52
Buff, Sheila, 165
Burgess, George, 80
Burritt, Elihu, 71
Burrowing Day-Owl, 115, 122pl.56, 123pl.57, 125

Caerulean Wood-Warbler, 76, 81
California Grey Squirrel, 108
Californian Turkey Vulture, 113, 115, 120pl.54, 121pl.55
Californian Vulture (in Townsend), 23pl.15
Camera lucida, 37, 51, 86pl.33; method of, 54
Camera Lucida, The, 66pl.29
Canada Jay, 50, 64pl.25-26, 65pl.27-28
Canadian Woodpecker, 77, 101
Canvasback, 138
Carleton, Lieutenant, 147pl.66
Carolina Paroquet, 11, 18pl.4
Cass, Lewis (Secretary of War), 31
Cassin, John: death of, 185n.22; involvement of, with Audubon books, 114-15, 125; proposal of, 109-12, 117pl.51; subscribes, 102
Catesby, Mark, 6, 7, 31, 39, 138
Catlin, George, 107, 131, 133, 137, 139, 144pl.63
Chapman, John G., 149
Charles X, king of France, 30
Chevalier, J. B.: as agent for JJA, 51, 70, 73, 80, 178n.22; and Bowen, 74; name of, on title page, 176n.12; problems with, 82-83; response of, to JWA's drawings, 74, 75; and Townsend, 23pl.15, 49
Children, John George, 16, 27, 28
Childs, Cephas G., 11

Index

Childs and Inman, 20pl.10, 31
Childs, J. W. (colorist), 81
Childs and Lehman, 28
Chromistes, 127
Chromolithography, 110, 125
Civil War, effect of, 127
Clark, William, 14
Clay, Henry, 14
Clinton, De Witt, 14
Col. Abert's Squirrel, 108
Cole, Thomas, 137, 139, 150
Collie's Squirrel, 108
Collins, John, 163, 178n.20
Colnaghi, Paul, 25
Coloring, quality of, 29, 62pl.23, 126-27, 172n46. *See also* Colorists
Colorado State University, 165
Colorists, 55, 56, 99pl.49
Common American Wild Cat, 147pl.66
Common Buzzard, 136
Common Grackle, 6
Common Mocking Bird, 81, 96pl.46, 97pl.47, 182n.20
Common Osprey, 130
Common Tern, 101
Comparisons among plates, 81-82, 84, 101, 113-14, 115-16, 125, 127, 129-30
Cooper, William, 171n.37
Cormorant, 136
Cornell University, 163, 164
C. P. (artist), 163
Craighead, R. (printer), 164
Crowninshield, Miss C. (subscriber), 176n.10, 182n.15
Cruickshank, Frederick, 19pl.5-8
Cuntz, Hermann F., 171n.37
Curled-Crested Phaleris, 77
Currier and Ives, 55, 56
Cuvier, Baron Georges, 27, 109

David, Jacques-Louis, 3, 168nn.4, 9, 169n.16
D'Avignon, Francis, 148pl.67, 149
De Bry, Theodore, 7

Degas, Edgar (mention), 102
Delorme, J. (artist), 23pl.15, 49
D'Obigny, Dr. Charles, 3
Dock, George, Jr., 165
Domestic Fowl (Bewick), 22pl.12
Dorsey, E. G., 49, 52, 55
Downy Woodpecker, 77, 101
Drake, Dr. Daniel, 4
Drummond, Thomas, 36, 173n.9
Duke of Orléans, 27
Dunlap, William, 32, 132
Dürer, Albrecht, 138
Dusky Squirrel, 108
Duval, Peter S., 52, 110, 178n.21

Elements of Botany (Barton), 6
Emerson, Ralph Waldo, 133, 135
Endicott, George, 52, 81, 182n.20
Engraving: decline of, 35, 45; processes referred to as, 26; technique of, 10-11
Estes and Lauriat, 128
Evening Post (New York), 60
Everett, Congressman Edward, 30

Fairman, Gideon, 11
Family Magazine, 149
Fauna Boreali-Americana (Richardson), 22pl.13
Fernandina, Count (of Cuba), 78
Fisher, S. Rhoads (Secretary of the Navy), 37
Fish Hawk (in Warren), 124pl.59
Flexner, James Thomas, 133
Fork-tailed Flycatcher, 61pl.22
14 Miles N.W. of Altar, 115, 123pl.58
French Royal Academy of Sciences, 27
Friedrich, Caspar David, 137

Gallery of Illustrious Americans (Brady), 148pl.67, 149
Gardiner, Rockwell, 171n.37
George IV, king of England, 16, 30
George R. Lockwood and Son, 129. *See also* Lockwood, George R.; Lockwood, Richard B.; Roe Lockwood and Son

Index

Gesner, Conrad, 7
Gifford, Euphemia, 12, 170n.24
Gifford, Richard, 170n.24
Gilmor, Robert, Jr., 28, 102
Godey's, 104
Godwin, Parke, 59, 76
Golden Eagle, 82, 136
Golden-Eye Duck, 150
Gordon, Alexander (brother-in-law), 12
Gordon, Ann (sister-in-law), 12
Goshawk, 84, 139
Gos Hawk, 82
Gould, A. A., 104
Gould, John, 7, 46, 48
Grammar of Botany, The (Smith), 52
Great American White Egret, 116
Great Blue Heron, 6, 39, 44pl.21, 129
Great Crow Blackbird, 10, 20pl.9
Greater Prairie Chicken, 17pl.2
Great-footed Hawk, 141pl.60. See also Peregrine falcons
Great Horned Owl, 130
Great Horned Owl (Swainson), 22pl.13
Great Piece of Turf, The (Dürer), 138
Great White Heron, 43pl.20, 45
Greenshank, 127
Grimshaw, James, 37

Hall, H. B., 163
Hall, James, 52, 60, 137. *See also History of the Indian Tribes . . .*
Harlan, Dr. Richard, 46
Harlan's Buzzard, 114
Harris, Edward, 32, 37, 50, 106
Harris' Finch, 77, 101
Harris's Buzzard, 114
Harris's Hawk, 176n.9
Harry Ransom Humanities Research Center (UT), 163, 164, 185n.19
Hartley, Marsden, 150
Havell, Henry, 73, 74, 81, 106
Havell, Robert, Jr.: contributions of, to plates, 26, 39, 40, 45, 50, 139; hiring of, 25; immigration of, 45, 48; need of, for assistants, 171n.36; relationship of, with JJA, 27, 28, 70; and royal octavo, 35, 102; skill of, 26, 172n.46. *See also* plates, pp. 41–44, 64–65, 94, 141–43, 145, 146
Havell, Robert, Sr., 16, 25
Heade, Martin Johnson, 132
Heermann, Dr. Louis and Mrs., 5
Hemlock Warbler, 76, 77, 81, 88pl.35, 89pl.36
Henslow, John Stevens, 30
Hermit Wood-Warbler, 77, 82, 90pl.37
Hill Memorial Library (LSU), 55, 76, 81, 113, 182n.20
Histoire de la Nature des Oyseaux (Belon), 7
Historia Animalium (Gesner), 7
History of British Birds, A (Yarrell), 23pl.14
History of the Indian Tribes of North America (McKenney and Hall), 52, 74, 80
History of the Rise and Progress of the Arts of Design in the United States (Dunlap), 32, 132
Hitchcock, William E., 55, 112, 114, 115, 163
Holley, Mary Austin, 36, 38
Homer, Winslow, 150
Hone, Philip (mayor of New York), 32
House-Sparrow, The (Yarrell), 23pl.14
Houston, Sam, 37-38
Howell, Warren, 170n.27
Hudson River School, 137
Hudsonian Godwit, 136
Hunter-Naturalist, The (Webber), 149
Hutchins's Barnacle Goose, 40

Illustrated books, 7
Illustrations of British Ornithology (Jameson and Selby), 14, 46
Indigo Bird, 50, 176n.9
Indigo Bunting, 176n.9. *See also Indigo Bird*
Inman, Henry, 52, 163
Ivory-Billed Woodpecker, 11
Ivory Gull, 77, 93pl.42, 101

Jackall Fox, 108
Jackson, Andrew, 14, 28, 169n.10

209

Index

James, Dr. Edwin, 8
Jameson, Robert, 14
Jardine, Sir William, 46
J. B. Lippincott, 111, 112
J. C. (artist), 163. *See also* plates, pp. 62, 91, 92, 95
John James Audubon (Cruickshank), 19pl.5
John James Audubon (D'Avignon), 148pl.67
Johnson, John Taylor, 128
Johnson Reprint Corporation, 167n.2
John Woodhouse Audubon (Cruickshank), 19pl.8
Jones, Dr. Thomas P., 27
Journal of the Franklin Institute, 27
Julius Bien and Company, 125. *See also* Bien, Julius

Kelly, Ellsworth, 150
Kidd, Joseph Bartholomew, 15-16, 25, 34, 40, 132
King, Charles Bird, 52
Kirtland, Jared P., 111
Knobbed-Billed Phaleris, 77

Landseer, Thomas, 35
Larkin, Oliver W., 133
Lawrence, Amos (subscriber), 182n.15
Lawrence, Sir Thomas, 16
Lawson, Alexander, 8, 9-11, 20pl.9, 21pl.11, 138
Lear, Edward, 46
Least Flycatcher, 56, 63pl.24, 84
Least Pewee Flycatcher, 84
Least Tern, 77
Legaré, J. D., 60
Lehman, George: assistance of, 40, 43pl.20, 45, 138, 143pl.62; hiring of, 28; traveling with JJA, 31
le Moyne, Jaques, 7
Lesser terns, 134
Lesueur, Charles-Alexandre, 11
Library of Congress, 28, 30
Linnaean Society, 16
Literary Gazette, 30

Lithographer (Prang), 67pl.30
Lithography: popularity of, 45; process of, 53-55, 67pl.30; and reprint of *Birds*, 126; and royal octavo, 20pl.10, 31, 34, 36, 51
Little and Brown, 71, 79, 83
Little Screech Owl, 28
Lizars, William H., 13, 14, 15, 16, 25-26, 45
Lockwood, George R., 164, 187n.36. *See also* George R. Lockwood and Son; Roe Lockwood and Son
Lockwood, Richard B., 129, 187n.36. *See also* George R. Lockwood and Son; Roe Lockwood and Son
Long-Billed Curlew, 31, 101, 138
Long, Major Stephen H., 4, 8, 36
Louisiana Hawk, 50
Louisiana Heron, 39, 138, 143pl.62
Louisiana Water Thrush, 41pl.17
Louisiana State University, 55, 76, 164. *See also* Hill Memorial Library
Lucy Bakewell Audubon (Cruickshank), 19pl.6

MacGillivray, William, 30, 35
Macmillan Company, 167n.2
Magazine of Natural History, 45
Mandeville, Bernard, 169n.10
Maria's Woodpecker. Three-toed Woodpecker. Phillips Woodpecker. Canadian Woodpecker. Harris's Woodpecker. Audubon's Woodpecker, 94pl.44
Marsh Hawk, 50, 176n.9
Marsh Hens, 20pl.10, 31, 51
Martin, H. Bradley, 167n.2
Martin, Maria, 31, 40, 61pl.22, 69
Mason, Joseph, 5, 6, 10, 17pl.3, 41pl.17, 59, 138
McKenney, Thomas, 52. *See also History of the Indian Tribes...*
Mease, Dr. James, 8
Mercury (Philadelphia), 59
Mexican Ground Squirrel, 108
Mexican Marmot Squirrel, 108
Mier y Terán, General Manuel de, 36
Miller, Alfred Jacob, 110

Index

Mississippi Kite, Tennessee Warbler, Kentucky Warbler and Prairie Warbler (in Wilson), 21pl.11
Missouri Meadow Lark, 77, 84, 100pl.50
Mocking Bird, 11, 136
Monthly Chronicle (London), 2
Morton, Dr. S. G., 47
Mount, William Sidney, 150
Mountain Brook Mink, 108
Mourning Dove, 138, 146pl.65
Mourning Ground Warbler, 75, 76, 85pl.32
Mumford, Lewis, 5
Musson, Germain, 102

Natural History of Carolina, Florida, and the Bahama Islands (Catesby), 6, 7
Naturalist's Library, 104
Neal, John, 59-60
Neill, Patrick, 14, 30
New England Galaxy, 59
Newman, Harry Shaw, 171n.37
New-York Historical Society, 128
New York Lyceum of Natural History, 11, 48
New York Mirror, 46
New York Times, 150
Noddy Tern, 40
Nolte, Vincent, 12
North American Review, 46
Northern Goshawk, 40
Novak, Barbara, 132
Nuttall, Thomas, 36

Ohio State University, 165
Old Print Shop, 171n.37
Ord, George, 9, 10, 11, 109, 169n.18
Ornithologia (Aldrovandi), 7
Ornithological Biography, 1, 58, 133, 137; publication of, 30, 35, 45; review of, 109; and royal octavo, 51; Texas observations in, 38
Ornithology of the United States of America . . . (Townsend), 23pl.15, 49
Osprey, 136
Otis, Bass, 52

Pacific Northwest, 36
Pacific Railroad Reports, 114, 125
Parkman, Dr. George, 140
Parley's Magazine, 104
Parsons, Charles, 163
Patriot (Baltimore), 60
Peale, Charles Willson, 6, 8, 138
Peale, Rembrandt, 8
Peale, Titian Ramsay, 8, 9
Penfold, Edmund (subscriber), 182n.15
Peregrine falcons, 135, 136. *See also Great-footed Hawk*
Peterson, Roger Tory, 165
Philadelphia Academy of Science, 36, 106
Philarète-Chasles (critic), 140
Phillips, Dr. Benjamin, 50
Phillips Woodpecker, 95pl.45
Philosophical Society of York, 16
Pirrie, Lucy, 6
Plumed partridge, 139
Prang, Louis, 67pl.30
Prince Maximilian of Wied-Neu Wied, 7, 107
Purple Grackle, 138

Quadrupeds. *See Viviparous Quadrupeds of North America*

Rabine, Jeanne (mother), 3
Rathbone family, 12, 16, 169n.16
Rathbone, Mrs. William, frontispiece, 13
Rathbone, Richard, 36
Rau, Edward H., 55
Red-Bellied Nuthatch, 77
Reddish Egret, 116
Red-Eyed Vireo, 77
Red-tailed Buzzard, 82
Redouté, Pierre Joseph, 27
Remington, Frederic, 170n.20
Report on the Birds of Pennsylvania (Warren), 124pl.59, 129-30
Rhett, Congressman R. Barnwell, 149
Rice University, 186n.23
Richardson, John, 22pl.13, 35

Index

Rider, Alexander, 10, 20pl.9, 169n.18
Right and Left (Homer), 150
Rocky Mountain Flycatcher, 84
Rocky Mountain Plover, 77
Roe Lockwood and Son, 115, 125. *See also* George R. Lockwood and Son; Lockwood, George R.; Lockwood, Richard B.
Romanticism, 134
Roscoe, William, 13
Roseate Spoonbill, 101, 138
Rosenthal, Lewis N., 110, 111
Rosenthal, Max, 125
Royal Society, 15, 16
Rozier, Ferdinand, 3, 4
Ruffed Grouse, 138, 145pl.64

Sandwich Tern, 77, 116
Sarony and Company, 178n.21
Saturday Courier (Philadelphia), 51, 60, 70, 140, 150; comments of, regarding royal octavo, 102, 104
Say, Thomas, 116
Scott, Sir Walter, 13, 15
Selby, Prideaux John, 7, 14, 46
Senefelder, Alois, 52
Series of Picturesque Views of Noblemen's and Gentlemen's Seats . . . , A (Havell, Sr.), 25
Series of Picturesque Views of the River Thames, A (Havell, Sr.), 25
Seymour, Samuel, 8
Shattuck, George C., Sr., 140
Sherwell, Amanda E. A. (colorist), 56
Short-eared Owl, 116
Short-legged Pewit Flycatcher, 84
Silliman's, 131
Sills, Joseph, 52
Sitwell, Sacheverell, 46, 150
Slender-Billed Guillemot, 77
Small-headed Flycatcher, 84
Smith, Sir James Edward, 52
Snowy Heron, 116, 138
Sora Rail, 77, 93pl.43
Sotheby's, 167n.2

Southern Cabinet (Charleston), 57, 59, 60, 104
Southern Methodist University, 164, 186n.23
Sparrow Falcon, 101
Spotted Sandpiper, 38, 138
Sprague, Isaac, 106
Stark Museum of Art, 49, 50, 56, 128, 164
Stephens, Henry Louis, 110
Stewart, Sir William Drummond, 106
Subscriptions: demographics of, for octavo, 101-03; for double elephant folio, 14-15, 27, 30, 46, 70; incomplete, 175n.25; practice of selling, 13-14, 111; for *Quadrupeds*, 108; for royal octavo, 58, 69-71, 75, 78, 182n.15
Sully, Thomas, 8, 9, 11, 16
Swainson's Swamp Warbler, 113, 118pl.52, 119pl.53
Swainson, William, 22pl.13, 35, 45, 48
Synopsis of the Birds of North America, 45, 50

Tallon, Mary (colorist), 56
Taylor, W. Thomas, 182n.16, 184n.7, 185n.18
Telegraph and Texas Register, 37
Texan Turtle Dove, 84
Texas A&M University, 164
Texas Memorial Museum, 185n.19
Texas Southern University, 164
Thielepape, Wilhelm C. A., 177n.15
Thoreau, Henry David, 114, 133, 135
Townsend, Dr. John Kirk, 23pl.15, 36, 48-49, 68pl.31; as sales agent for JJA, 52, 177n.12
Townsend's Sandpiper, 50, 176n.9
Townsend's SurfBird, 176n.9. *See also Townsend's Sandpiper*
Townsend's Warbler, 39
Townsend's Wood Warbler, 77
Transco Energy Company, 177n.13
Trembley, R. (artist), 55, 75, 76, 163. *See also* plates, pp. 61, 63, 65, 85, 88, 98, 99, 118, 120, 121, 122
Tropic Bird, 39
Tuckerman, Henry, 132
Tufted Puffin, 77
Tufted titmouse, 134

212

Index

Turkey Buzzard, 40
Turner, C. (engraver), 19

University of Edinburgh, 167n.2
University of Michigan, 164
University of Minnesota Bio-Medical Library, 164
University of Texas, 163, 164
U.S. Department of Agriculture, 79
U.S. Secretary of State, 114

Vanderlyn, John, 169n.10
Victor Gifford Audubon (Cruickshank), 19pl.7
Virginian Partridge, 136
Virginian Rail, 77
Viviparous Quadrupeds of North America, 47, 82, 106-08; octavo edition of, 108, 114, 185n.20; paintings for, 76, 84, 108; production of, 107; publication of, 83; sales of, 108; subscribers to, 108

W. (artist), 75, 86pl.33, 178n.20, 181n.4
Ward, Henry (taxidermist), 31
Warren, B. H., 124pl.59, 129, 130
Waterton, Charles, 169n.18
Watson, John Frampton, 178n.20
Webber, Charles W., 110, 149
Webster, Senator Daniel, 71
Welcome Partridge, 77, 82, 92pl.40-41
Werner, Abraham G., 170n.27

Wernerian Society of Natural History, 13, 15, 170n.27
Western duck, 139
Western Monthly Magazine, 60, 137
Western Museum (Cincinnati), 4
Western Shore-Lark, 77
White, George Gorgas, 111
White-Headed Sea Eagle, or Bald Eagle, 82, 114, 115
White-legged oyster catcher, 139
Whooping Crane, 38, 77, 82
Wild Turkey, 11, 14, 117, 126, 127
Willow Ptarmigan, 77, 91pl.38-39, 101
Wilmerding, John, 133
Wilson, Alexander, 3, 8, 9, 11, 21pl.11
Wilson's Flycatching-Warbler, 76, 77, 81
Winterfield, Charles, 10, 140
Wolf, Josef, 46
Wollaston, William Hyde, 54
Woodhouse, Elizabeth, 170n.24
Wood Ibiss, 39
Woodside, Abraham, 178n.20
Wood thrush, 134
Wyeth, Nathaniel, 36

Yale University, 164, 182n.15
Yarrell, William, 23pl.14, 48, 104
Yellow-Billed Cuckoo, 6, 26, 77, 101, 138
Yellow-Bellied Flycatcher, 77
Yellow-Crown Warbler, 40

This book was produced in an edition of 1500 copies, designed by W. Thomas Taylor. The type is Monotype Bembo, cast and printed by Bradley Hutchinson, on Mohawk Superfine paper. Page make-up and hand composition is by Neil Furqueron. The calligraphy is by Jerry Kelly. The illustrations were printed by David Holman at the Wind River Press, and the binding was executed at Custom Bookbinders.